Managing Horses
on Small Properties

Managing Horses on Small Properties

Jane Myers

National Library of Australia Cataloguing-in-Publication entry

Myers, Jane.
Managing horses on small properties.
ISBN 0 643 09067 3.
1. Horses – Australia. 2. Farms, Small – Australia – Management. I. Title.
636.100994

Available from
Landlinks Press
150 Oxford Street (PO Box 1139)
Collingwood VIC 3066
Australia

Telephone: +61 3 9662 7666
Local call: 1300 788 000 (Australia only)
Fax: +61 3 9662 7555
Email: publishing.sales@csiro.au
Web site: www.landlinks.com

Front cover
Photo by istockphoto
Back cover
Photo by Jane Myers

Set in Minion 10.5/13
Cover design by James Kelly
Typeset by Paul Dickenson
Index by Russell Brooks
Printed in Australia by BPA Print Group

Acknowledgments

Stuart Myers, Queensland
Annie Minton, Horses and People Photography, Queensland
Sally Brunton, Aranui Stud, Queensland
Todd Cunningham, Equestrian Land Developments, Queensland
Julie Moran, Victoria
Margaret Blanchard, Victoria
The Horse Shed Shop, Victoria
Geoff Kerrison, Queensland
University of Queensland Gatton Campus
Dale Sutten, Wadham Park, Queensland
Don Baxter, NSW
West Rock Farm, Victoria
Richard Mullen, Gallagher Australia
Natalie Waters, Victoria
Logan Village Fencing and Produce, Queensland
Yves Cousinard, Pacific Performance Horses, Queensland
Belcam International, Queensland

Contents

1

Introduction

In the last century it was predicted that the number of horses would decline as horses were increasingly being superseded by motorised vehicles. What was not taken into account was the increase in disposable income coupled with a rise in expectations for a better lifestyle. People now strive to enjoy their recreation time more than ever before and this has led to horses fulfilling a new role in human society. Even people who have no wish to own one, or ride, get pleasure from looking at a horse. In Australia and New Zealand horses play a large part in many people's lives and horse ownership is on the increase.

Although horses are now largely used for recreation purposes, they contribute a huge amount to the economy of both Australia and New Zealand. In Australia, for example, they contribute an estimated $6.2 billion each year. This makes the horse industry one of the largest primary industries. There are now more horses in use than when they were used mainly as a work animal.

For New Zealanders and Australians it is relatively easy for the dream of horse ownership to come true because the cost of horses and land is cheaper in comparison with many other countries. Many make a deliberate lifestyle choice to buy land and keep horses in preference to agisting their horses on someone else's property.

However, with horse ownership come many responsibilities including caring for the horse and the land on which the horse lives. Unlike a machine, the horse must be cared for on a daily basis whether or not it is being used that day – although for many people caring for horses is not a chore but is a large part of the pleasure of owning a horse. Horses thrive in an environment that is as close to natural as possible. This means that they have access to pasture and clean water, the companionship of other horses, shade and shelter. Their basic requirements must be met in order for them to thrive both physically and mentally. Regular foot care, teeth care and worming are also essential practices of horse care.

Caring for the land can be equally rewarding once you have the requisite knowledge and skills. Caring for your property is as important as caring for your horse. Horses are

Old horse stalls Credit: Jane Myers

not native to Australia or New Zealand. This means that horses have a huge impact on the environment if they are not managed well.

Property management includes how the pastures and facilities are managed, how to deal with water and with waste and the management of horses and their impact on the environment. Good property management has many benefits including protecting the health and wellbeing of your horses, the water, the land and wildlife. It also increases the value of your property and it keeps your neighbours happy. It does not need to be an expensive undertaking. In fact often a slight change in operations can lead to big savings, for example improved pasture equals smaller feed bills, reduced mud or dust mean better horse health and fewer vet bills, better manure management results in a liability becoming an asset.

It is possible to create property management systems that, once established, are easy to run both in terms of time and expense and at the same time are less damaging to or even enhance the environment. For example, pasture that is grown for horses protects the soil, trees that are grown for shelter or even fodder also provide habitat for birds. Good land care involves such practices as rotating horses around paddocks so that pasture has time to recover, harrowing larger paddocks to spread manure and fertilising when necessary. A better-managed property provides a variety of plants that in turn supports many animals – an ecosystem. The more diverse the ecosystem, the better it can cope with environmental disasters such as floods, drought, pests and disease.

Also associated with environmental matters – it is not widely known that horse owners are already heavily involved in recycling. The horse industry is one of the largest end users of recycled tyre products including rubber stall and float mats, tyre feeders and shredded rubber arena surfaces. Other recycled products that are utilised on a large scale in the horse industry include feed containers made from steel and PVC drums. Horses also provide many tonnes of manure that are used by both private and commercial gardeners.

Many people who have chosen to live on a small property with their horses want to do the right thing but do not know where to start. This book will be helpful to property owners who are interested in combining healthy horse management and good land management. The aim is to give you options, resources and references that you can implement so that your horse care and land care systems will result in a better lifestyle for yourselves, your horses and your neighbours as well as benefiting the environment.

Jane Myers
January 2005

2

Horse management

This section of the book describes the basics of horse care and welfare especially as it relates to small properties. Keeping horses in small areas tends to raise issues that may not be a problem on larger properties if more pasture is available. When horses are kept on smaller areas their management becomes more intensive, many things have to be taken into account if the horses are to be cared for properly and, at the same time, land degradation kept to a minimum.

Natural horse behaviour

To care for horses and land properly a horse owner must understand horse behaviour. This is so that horses are not stressed and can live as close to natural a life as possible.

Horse characteristics

In all interactions with horses, it must be remembered that they are 'fright and flight' animals. This means that in the wild they would be the prey and that their first reaction to any threat or source of fright is to run away. Everything about the horse, the way it is built and the way it behaves, is designed to meet this primary instinct.

The horse is a herbivore and part of the food chain in its natural environment. Being a herd animal, it does not live alone by choice. Knowing these facts helps us to understand why the horse behaves the way it does.

Wild or feral horses are not forced to graze over their manure due to the area that they have available. Also, in the wild, many species graze the same areas, each eating around each other's manure and taking in each other's parasitic worms, doing this kills these worms. Even horses that live at pasture in captivity do not live the same life as their wild cousins as wild equines cover many miles a day to forage, drink and find shelter.

Domestic horses, feral horses and their wild cousins (zebras and asses) have many behavioural characteristics in common. Their behaviour is governed by innate instincts

Figure 2.1 The fright and flight instinct – running away Credit: Julie Moran

that drive them to behave a certain way, so we should not be surprised if horses sometimes do things that seem very strange to us. Learning about their natural behaviour can help us to understand why they do the things they do and how to manage them better. This leads to a calmer horse and happier owner.

Body language

Horses use body language to communicate with each other. When working around horses, an understanding of their body language is essential for safety and to be able to provide appropriate training and management techniques. If the body language of the horse is interpreted correctly, the handler or rider can act to forestall problems.

An observant rider or handler should be able to identify the outward signs a horse exhibits that indicate what it is likely to do next. The ears are a good barometer as they follow the movement of the eyes. When riding, it is possible to tell what the horse is concentrating on by looking at its ears.

When a horse is frightened it shows particularly strong body language. Because it is in high alert it usually freezes. This is because being still allows the horse to see and hear the danger while being less noticeable to the potential predator. The head will be high and the horse will be staring and have its ears pricked in the direction of whatever is holding its attention. It may also emit a loud snort. Then the horse may then spin around and try to run away.

Horses often give many signals before they act, such as laying back their ears before biting or kicking. It is up to us to learn about this body language so that we can move around horses safely.

Be careful when in confined spaces with horses. The reason for this is that although horses prefer to move away from threats, they are forced to defend themselves if they cannot get away.

Figure 2.2 On the alert Credit: Julie Moran

Eating behaviour

The horse is designed to eat almost continually, consuming roughage for up to 20 hours per day. The horse has huge muscles in the jaw, which enable it to chew for hours at a time. When feeding horses it is best to remember that the horse has evolved to eat large amounts of low energy fibrous food. This type of food takes a long time to chew and digest and keeps the horse occupied for most of the day. Horses have a primal need to chew since this is what they are designed to do. Chewing is also important for the production of saliva that buffers the acid in the stomach.

Unlike a ruminant (such as a cow or sheep), a horse does not need to rest and ruminate to process and digest food. On the contrary, a horse will graze almost continuously with fermentation being carried on in the hindgut as the horse is grazing. This means that a horse is on its feet for most of the time, ready to flee, and not weighed down by unmanageable quantities of forage.

It is quite natural for the horse to store body fat when the opportunity arises. This is due to an innate compulsion that drives the horse to eat more than is needed when food is available and store it as body fat. In the free-ranging horse, when grass is not so readily available (such as in the winter), these reserves of body fat can be utilised. Due to this natural behaviour the domestic horse, if it is not in work, can soon put on too much weight when grazing rich green grass. This can lead to a serious disease called laminitis (see p. 36), which is an extremely painful and sometimes life-threatening disease.

Modern day horse management has resulted in many horses being confined and fed meals that are high in energy but low in roughage. This results in long periods of time where the horse has nothing to do as these types of feed are eaten much more quickly (due to being more energy dense). This can cause problems, either behavioural (see p. 10), physiological or both.

One of the most important physiological problems that can occur when the gut is empty for too long is colic. Colic is often the result of the gut not being able to cope with the feeding regime (see p. 35).

Grazing and paddock behaviour

The way that horses behave affects pasture management. When grazing in confined areas (paddocks), horses will tend to dung in one area and graze in another. Horses will not eat around their own dung (this is thought to be a natural parasitic worm prevention strategy). Therefore, if left to their own devices, this behaviour leads to increasingly large areas where the grass gets long and rank (these areas are called 'roughs'), and decreasing areas that get overgrazed (these areas are called 'lawns'). Paddocks that have marked areas of roughs and lawns are often termed 'horse sick'.

This behaviour leads to less grazing available to the horse each year. The roughs over time also become dominated by weeds and the whole paddock has an imbalance of nutrients as the horse takes from the lawns (by grazing) and deposits (manure) in the roughs. This manure is high in potassium, phosphorus and magnesium, therefore the roughs receive lots of fertiliser from manure and the lawns do not.

In addition to their dunging behaviour, horses are highly selective grazers. They are able to be much more selective than cows because of their prehensile top lip which allows them to 'sift' what they want from what they do not want. Because they bite with their incisors (unlike cows which can only eat longer grass that they can wrap their tongue around) they can bite grass down to ground level. Horses, if left on a paddock for long enough, will select plants that they like and graze them to the ground and leave the rest to go rank. The types that they like will eventually get grazed out.

Horses tend to walk a lot when on pasture. This creates paths and bare areas because in their natural state horses travel many miles a day when grazing. Horses are constantly on the move when they graze and move from water to grassed areas. This can be a good deal of traffic for your paddocks.

Other behaviours that can be detrimental to paddocks are that horses will occasionally canter or gallop around in them. This depends on how much horses are confined before being turned out, how fit they are and how much energy they are receiving from their feed. Horses also like to create areas for rolling in (particularly in sandy soil).

Horses that are being fed supplementary feed at pasture will tend to stand around and wait, long before feed time. This can cause problems such as soil compaction near the entrance and exit to the paddock. Sometimes horses do not need the supplementary feed but get into the habit of standing around anyway.

Horses at pasture will graze for between 12 and 20 hours a day depending on the quality and quantity of grass (see p. 8). This is usually spread out throughout the day and night and is interspersed with sleeping and loafing (see p. 9). If a horse is removed from pasture for a period and then returned (such as when a horse is removed from the paddock for exercise), it will graze for a longer period without stopping to make up for the shortfall. For this reason, if removing horses from pasture for the sake of conserving pasture or to reduce the horse's intake (such as with a potential laminitis case), the horse will need to be removed for more than four hours, and up 12 if the pasture is good, to make a difference to the horse's total daily intake (see p. 111).

Figure 2.3 A horse sick paddock

Credit: Jane Myers

Careful management will reduce or eliminate the effects of these behaviours. Paddock management practices (see p. 99) in combination with periods of confinement will result in evenly grazed paddocks and reduced land degradation.

Herd instincts and social hierarchy

Horses are highly social animals that need other horses for company. This is because they are herd animals and never live alone by choice. Horses live in groups because in the wild there is safety in numbers. More sets of eyes and ears mean that predators can be seen sooner. Even when horses live in captivity these strong instincts are still very much a part of the horse's make-up and cannot be disregarded.

In the herd each horse has a place in the social hierarchy, sometimes called the 'pecking order'. This pecking order is not necessarily linear. Horses in captivity tend to demonstrate this pecking order more often because the way in which they are kept leads to this. For example, in the wild, grass is usually either available to all of the horses or none of them. In captivity we tend to feed horses which leads to much competition. We initiate pecking order aggression when we feed horses in the paddock together, in the wild no one comes along with a bucket to create this aggression.

Even horses that are at the bottom of the pecking order will still chose to stay with other horses rather than be alone. In fact two horses can appear to have a strong dislike for each other, but may still become frantic when separated.

The pecking order must be taken into account when a new horse is turned out into a paddock with strange horses, the new horse must be introduced gradually (see p. 16). The group must be watched carefully during these times to ensure that the horses do not injure each other.

Time budgets and social behaviour

Many animals have been studied in the wild state to discover what the 'time budget' is for that particular species. The time budget is the amount of time an animal spends

Figure 2.4 Bare areas in a horse paddock Credit: Jane Myers

doing the things it has to do throughout the day. The time budget of most predators involves short periods of high activity, to catch and eat prey, and long periods of inactivity, i.e. sleeping, to digest the prey they have eaten. Herbivores are very different. As well as having to be alert most of the time (watching and listening for predators), they have to eat for most of the day. Compared to a meat eater their food is low in calories and takes a long time to chew and digest. Studies have shown that the time budget of feral/wild horses is:

Grazing – between 12 and 20 hours a day
Sleeping – between 2 and 6 hours a day
Loafing – between 2 and 6 hours a day

Time spent grazing

The length of time spent grazing depends on the quality of the grass available. On better quality grass the horse will spend less time grazing and more time sleeping and loafing. When the grass is poor, such as in a drought, the horse will increase the total grazing time from 12 up to as much as 20 hours per day. The total time spent grazing is usually spread out with bouts of sleeping and loafing being interspersed throughout the day and night.

Time spent sleeping

Adult horses usually sleep for approximately four hours per day. Roughly two hours are spent sleeping lying down and two hours are spent sleeping standing up. Due to the large thorax of the horse, it actually uses less energy to sleep standing than lying down. Lying down rests the legs but the lungs have to work hard when the horse is stretched out on its side. This is why a horse often makes a groaning noise when prone, as breathing is quite an effort in this position.

This ability to sleep standing is unique to equines. They use what is termed 'the stay apparatus' to lock the joints in place. A horse can even rest one hind leg as it nods off to sleep. This behaviour allows a horse to move off quickly in an emergency.

Figure 2.5 Horses are herd animals Credit: Jane Myers

Notice that in a group of horses one horse usually stays standing when the others are asleep on the ground. This horse is looking out for danger (even if it is drowsing) while the others sleep more deeply. This is a good example of how herds operate. Horses that live alone do not get to benefit from this system of shared responsibility. Some nervous horses are even unwilling to lie down when they are on their own as they will have no other horse to watch over them as they sleep.

Time spent loafing

'Loafing' is a term that is used to group all the other things that horses do with their day. It includes such activities as mutual grooming, playing and standing in the shade, head to tail, swishing at flies – a common behaviour of horses especially in hot weather. In cold, wet weather horses will stand in a sheltered spot together to conserve heat.

Figure 2.6 Horses sleeping Credit: Jane Myers

Figure 2.7 Mutual grooming Credit: Jane Myers

Mutual grooming is where two horses approach each other and use their incisor teeth to 'groom' one another. Grooming is a very important behaviour for horses as it is a way of maintaining bonds between group members.

Playing is very important, especially in young horses, as it is an opportunity to learn some of the skills required for adult life.

Welfare issues

Good horse keeping involves acknowledging what is important to a horse. The basic needs of a horse are food and water, companionship, shelter from the elements and freedom to move around. When you take on the task of horse ownership you are responsible for ensuring that these basic needs are met. If they are not, the horse will be stressed. Some signs that indicate a horse is stressed include failure to thrive resulting in poor condition, poor performance and intractable behaviour.

Abnormal behaviour

There are certain behavioural problems associated with domesticated horses that are termed 'vices'. Vices are 'stereotypes' or obsessively repeated actions. This means that the horse will repeat an action, such as rocking from one front leg to the other (weaving) over and over again. Horses in the wild do not exhibit these behaviours because they are busy doing the things that they need to do to survive. Similar stereotypes are seen in some confined zoo animals. These abnormal behaviours arise because the animal is prevented from carrying out normal behaviour. Normal behaviour for a horse includes eating a diet that is very high in fibre (roughage), moving around continuously and socialising with other horses.

Unfortunately, in the horse world stereotypes are labelled 'vices' which implies that the horse is behaving badly. Some stereotypic behaviour is so entrenched in an animal that even when the environment is changed for the better the animal still performs the behaviour. However, over a long period of time, the behaviour usually becomes less frequent and the animal may only revert to it when stressed or excited.

Some stereotypes are oral in nature (i.e. those that are related to eating) and some are locomotive (i.e. those that are related to movement). Oral stereotypes include wind sucking/crib-biting, wood chewing and flank biting. Locomotive stereotypes include kicking stable walls, weaving, box walking and pacing up and down fence lines.

There are many myths connected to stereotypic behaviour, for example it is commonly believed that if another horse watches a horse performing stereotypic behaviour, it will also learn to do it. This is not true, as horses do not learn in this way. Isolating horses because of this belief will simply lead to increased stress in the affected horse and an increase in the behaviour.

On the market you can find many products that physically prevent horses from carrying out certain 'vices'. However, these products (such as wind-sucking collars) do not treat the cause of the problem and the horse will actually become more stressed. If your horse performs stereotypic behaviour try to think of ways in which you can improve that particular horse's lifestyle rather than simply prevent the horse from carrying out the behaviour. All horses benefit from the strategies shown below.

- An increase in the time spent grazing. Horses should be allowed to graze as much as possible. There is no better feed for a horse. Grass can be supplemented if the horse is in hard work.
- An increase in roughage in the diet. If you do not have grazing available, feed as much hay as possible. This means that the horse spends more time chewing which is what it is designed to do.
- An increase in the time spent with other horses. This means physically being with another horse, not just on the other side of the fence from one, which is far more dangerous due to the risk of fence injuries. Horses will hang around the fence if they cannot be together and will not go off and graze.

See Chapter 12 for recommended reading on the subject of horse behaviour.

Safe horse management

Horse-related activities are widely acknowledged as being potentially one of the most dangerous leisure pursuits. However, the level of risk can be minimised by addressing risk management strategies, that is, assessing the various risks involved and planning to minimise the occurrence of accidents.

Remember that even the best-trained horse is still a horse, especially when it is stressed. In some respects, the most dangerous horses are those that are regarded as quiet because it is easy to become complacent with them.

Horses must be handled firmly and consistently; any rewards or punishment (punishing a horse is not usually necessary) must be given immediately otherwise the horse cannot connect the two.

Even the quietest horse can react very quickly if it is frightened. Its first choice is to run, and if it is not able to do this it will defend itself. This is because the horse has evolved to have flight as its primary response to threat (unlike many other grazing animals). Standing up for itself is a secondary choice that it will implement only if it cannot run away.

Correct handling involves learning a general set of rules that should be applied to all horses, keeping in mind that each horse is an individual.

Remember that in order of importance the safety of humans is paramount, horses come second. You can replace a horse but not a family member or friend.

Minimum training requirements

There are certain minimum training requirements for horses to ensure their safety in various situations. All domestic horses (including young horses) should be able to be caught and led, tied up, handled and loaded onto a float. In the event of an impending fire or flood they can then be moved if necessary. In an emergency situation this may have to be carried out by someone that the horse is not used to, so the horse should be trained to a level that ensures that the horse will respond correctly for anyone, not just its owner. A horse that is trained to yield to pressure (and will therefore lead anywhere at anytime) rather than resist pressure is safer to handle and less likely to cause injury to people or itself.

Moving around horses

A horse should be either tied or held if it is not in a stable, yard or paddock. If you are handling a horse, it should be wearing a headcollar and lead rope as a minimum: never work around horses that are unrestrained. Do not clean a stable or a yard with a loose horse in it. Remove and tie the horse elsewhere while you clean the area, this is also better for the health of the horse as it will reduce the amount of fumes and dust inhaled by the horse (see p. 20). Be careful whenever picking up manure near horses as this usually involves bending down.

When moving around horses, the safest place to be is either very close or far enough away to be out of range of a bite or kick, which is approximately three to four metres away (bear in mind that a horse can run backwards and kick at the same time). The greatest impact from a kick is at its extremity. If you walk behind a horse at arm's length and it kicks out, by the time the leg contacts your body it will be at full speed.

If you are in close contact with the horse and it decides to kick, the kick will be more of a shove. If you are touching the horse you will feel the horse tense muscles or shift weight, which may be a sign that it is about to kick. If you suspect that a horse is about to kick, move towards the shoulder not further out behind where it will be able to reach you.

By staying close or in contact with the horse and speaking to it, the horse will know where you are and will be less likely to be startled by a sudden movement. If you have to walk behind the horse at close proximity keep your arm across your chest to protect your ribs. Keep talking to the horse so that it knows where you are; remember it cannot see you unless its head is high. Never walk behind a horse that has its head down eating. Never stand directly behind a horse (even when washing the tail) or directly in front. As well as being dangerous spots they are the areas that the horse can see the least.

Having said all of this, stay out of kicking range wherever possible. The shoulder end of the horse is usually safer than the back end. When accustomed to the body language of horses you will be able to read the warning signs that horses invariably give before biting or kicking, such as laying the ears back or picking up a leg. Until that time be extra observant when moving around horses.

A safe horse area

Your horse area and paddocks should be safe for both people and horses to use. Many accidents occur in these areas that could be prevented.

Stable buildings in particular are prone to fire because of the materials usually kept within. Always equip your stable yard with the following:

- an alarm system that is regularly tested
- suitable fire extinguishers, also regularly tested
- a hose permanently attached to a tap, with sufficient length and pressure to reach all buildings
- 'No Smoking' signs that are prominently displayed.

No flammable material such as petrol, kerosene or paint should be stored near the stable area and on no account should refuse be burnt close to the stables. Large quantities of hay should not be stored in or adjacent to stables.

Never walk a horse over the top of an electrical extension lead; leads should never be on the floor. Do not leave extension cables lying in hay or shavings. All coiled electrical cords generate heat when in use and there is a very real possibility of fire caused by short-circuiting.

Keep stables free from cobwebs. When cobwebs become covered in dust they make the flashpoint lower.

Electric sockets should be horse safe. They should be placed out of the way of inquisitive mouths. Make a monthly check for loose wires in plug tops, and for cracked plugs. Check for signs of frayed, cracked or chewed (could be due to rats) leads. Also check for signs of overheating on plug tops and socket outlets.

Many accidents are caused because things are left lying around. Every item should have a home and should be put away as soon as it is no longer needed. This includes shovels, rakes, brooms and buckets etc. Hoses should be hung rather than left lying on the ground.

Doors and gates should either swing both ways, in the case of paddock gates, or should open outwards in the case of yard gates and stable doors (sliding doors are even better). Doors and gates should be kept closed when not in use.

Headcollars should be hung up whenever they are not on a horse. This includes when the lead is still tied to a post because the horse is being worked and so on. A person or horse can be tripped up by a hanging headcollar. Always leave headcollars in one spot so that you will be able to find them quickly and easily in an emergency, even if it is dark.

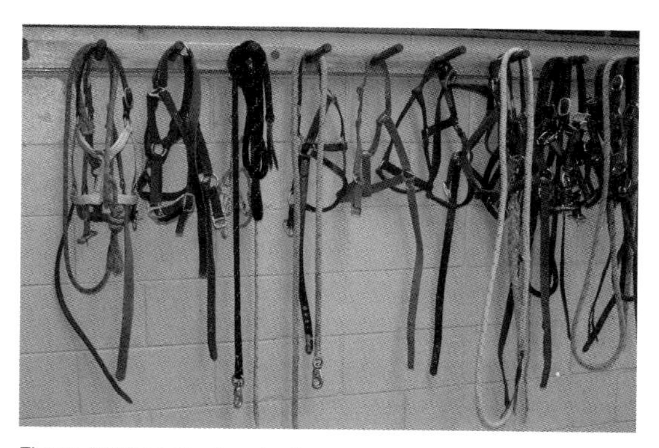

Figure 2.8 Hanging headcollars Credit: Annie Minton

Make sure there are no implements left in the paddock such as harrows – these can quickly get covered in grass and can cause fatal accidents. Fence wire left around, holes from rabbits, weeds, stakes sticking up, projections on fences, branches with low projections are all potential causes of accidents.

See Chapter 12 for resources on preventing and protecting your property from bushfire.

Sensitive areas

The most sensitive areas on the body of the horse include the nose, ears, feet, legs, and flanks. The nose is the area that predators grab when pulling down a horse so horses are naturally wary of anything touching this area. Also the nose has many nerve endings making this area very sensitive. Due to the positioning of the eyes and not being able to tilt the head in an upward direction beyond a certain point, the horse cannot see directly above its head. The ears, due to being right on the top of the head, are used by the horse (as well as to hear with) to feel if something is situated above it. This is why a horse will sometimes panic if something grabs the ears, or if the ears brush on a roof that is too low for example. Injury to the feet and legs are detrimental to the survival of a wild horse, so the horse is naturally wary about allowing you to handle them at first. The flanks are an erogenous zone and are therefore very ticklish, especially on mares; this area is gently sniffed by the stallion to test a mare's receptiveness to him.

Children and horses

Small children should be kept under control at all times around horses. Children can get excited when they see horses but they do not know the dangers. Horses do not always recognise children for what they are (small humans) and can get frightened or defensive if they are not used to them.

Never leave children in prams in a horse area. The child is in a very dangerous position and a horse will not initially associate a pram with being harmless.

Dogs and horses

Dogs that are not used to horses can get very excited or frightened by them and quickly start to make a nuisance of themselves, through barking or growling. Horses that are not used to dogs can get alarmed – after all a dog is a natural predator of a horse. Some horses are very defensive with dogs and will bite or strike with a front leg or kick with a hind leg.

Horses and dogs should be allowed to become accustomed to one another without either animal feeling threatened. If you own a horse that is frightened of dogs you need to improve this behaviour before riding out on the roads. Dogs will often run to the front of a property and will sometimes come out on to the road to bark or even chase a horse. This is potentially a very dangerous situation especially if you are riding a horse that is nervous about dogs.

Approaching a horse

Due to the restricted vision of the horse (it cannot see straight in front of its head nor directly behind its body), always approach at an angle, never directly from the front or the rear. Speak to the horse as you approach so that it is not startled. Look for the movement of an ear towards you that will signify that the horse is aware of your presence. Walk towards the horse in a smooth fashion, don't walk in awkward stops and starts or creep towards it. The first contact with the horse should be on the shoulder or neck, rubbing rather than patting (horses do not really appreciate being patted). Place an arm on either side of the neck and then slip the lead rope over the neck to secure the horse with the lead rope. Then put the headcollar or halter on, moving it slowly over the horse's nose and ears.

The procedure for approaching a horse is the same whether it is in a stable, yard or paddock. However, if the horse is facing away from the yard or stable entrance, it must be made to 'face up' before the handler enters the area. Never enter an enclosed area with a horse when it has its back to you. The horse can swing around before it is secured and either knock over or kick the handler.

Leading a horse

When leading through gateways and doorways stop the horse so that you can go through first. Always open doorways and gateways fully so that the horse does not bang its hips as it goes through. Openings should open outwards to avoid the horse getting trapped in them.

There are occasions when you may have to let a horse that you are leading go. This could be in a situation where you would otherwise be dragged. Hanging on to a horse in such an event will actually make things worse, and you will be injured. Avoid being in a situation where you have to lead a horse in a potentially unsafe area (for example on a road), if you are not experienced enough to handle the horse if it starts to get excited or frightened.

Tying up a horse

When tying up a horse, you need to select the area carefully and use good quality equipment. Many injuries occur to people and horses simply because of inexperience. It is difficult to understand how dangerous a horse that 'pulls back' is until you have seen it happen. If a horse is pulling back, it is panicking and is therefore very dangerous. If the horse is tied incorrectly then the danger is increased. There are some general rules when tying up horses that should always be followed:

Figure 2.9 A horse tied at the correct height, using a quick-release knot and a loop of twine between the rope and the tie spot
Credit: Jane Myers

- The horse should always be tied up in a safe place. A safe place is somewhere that has a non-slip surface, without obstacles and is not near other horses.
- The horse should be tied to something strong and secure. A solid post is usually the best place. Do not tie to the rail of a fence as the horse can pull it off. Never tie to the wires of a wire fence because if the horse pulls back it will pull the whole fence with it. This can happen and is usually catastrophic when it does.
- Tie the horse no lower than level with its eye, using a quick-release knot so that the horse can be released quickly in an emergency. Horses tied lower than their head height have more leverage if they pull back. Give the horse approximately 50 cm of rope between the head and the tie spot. If the rope is too long the horse may get a leg over the rope.
- Never leave a tied-up horse unless it is experienced at being tied and is not likely to pull back.
- If the horse has a history of pulling back or you do not know the horse, use a single loop of twine as link between the rope and the object that the horse is tied to. This twine should break in an emergency.
- Never tie up with the bridle reins. Use a headcollar and rope, and always untie the horse before removing the headcollar.
- Do not tie up with a rope halter. If the horse pulls back, damage can be done to the area behind the poll due to the narrow rope.
- Don't tie a horse to a float unless the float is attached to a car or truck. Do not tie up to a single float even if it is attached to a vehicle as a horse can pull it over.
- Have a knife handy just in case you have to quickly release the horse. In this case the rope or twine can be cut quickly if necessary.
- Keep your fingers out of the loops when tying a horse up. People have lost fingers in this situation when a horse has unexpectedly pulled back.
- Beware of tying a horse to the cross-rail on a hitching rail unless it is exceptionally well made (and many are not), because if the horse pulls back and the rail gives it can hit the horse and any handlers at full speed.

Releasing a horse

When releasing a horse into a paddock, walk it through the gate and pull the gate shut (but not fastened) behind you. Then turn the horse around so it faces the gate before you let it go, then you should step back out of the paddock immediately. Some horses kick out in play as they start to gallop off. You could be badly injured if you release the horse facing into the paddock. Turning the horse around gives you a fraction more time to get to safety.

When releasing a horse into a stable or enclosed yard, always turn the horse back towards the entrance so that the horse can be released without the handler having to walk past the back end of the horse.

Introducing a new horse

There are many ways that a new horse can be introduced to a group of horses. Remember that the existing group will have a pecking order and the introduction of a new member will disrupt this. It is not safe for anyone concerned to simply turn the

Figure 2.10 Introducing a new horse Credit: Jane Myers

new member out into the group in a small paddock. In a small space the new horse can be run into a fence by the other horses. It is better to let the new horse get to know the quietest member of the group by being turned out together and then, when these two horses are accustomed to each other, other group members can be added one at a time.

Routine horse care

Caring for horses takes a lot of commitment. Even horses that live in paddocks must be checked on a regular basis (daily as an absolute minimum). Unless you have limitless grazing, which is very unlikely on a small property, you will need to adhere to a system of management that involves periods spent in the stables/yards and periods spent grazing. These periods will alter in time depending on such factors as the weather and the amount of grass available. Horses that are kept fully stabled or yarded or are kept on a combined pasture/confined system will therefore need attention at least twice a day.

Horse care systems

To live in a large, grassy paddock with other horses, a good water supply and good shelter would be ideal for most horses. Unfortunately it is not always possible to provide this situation. However, by acknowledging that this is the ideal, it is usually possible to make changes to the living conditions of a horse that will enhance its lifestyle.

Horses can live in a variety of situations including a paddock, stable, yard or a combination of these situations and will adapt well as long as the owner can adhere to the basic principles of horse care.

Paddocked horses

If acreage is available, keeping horses in paddocks is less time consuming for the owner and results in a better lifestyle for the horse. Grass is a relatively cheap form of feed and it is what the horse evolved to eat. Grass provides the roughage that is essential to the wellbeing of the horse.

Horses in paddocks must be managed in such a way that they do not cause land degradation (see p. 90) which in addition to causing problems with the environment will result in progressively less available grass each year.

Horses are more settled if they have companions because they are first and foremost a herd animal (see p. 7) Studies have shown that horses eat more if they can see another horse. Whenever possible, horses should be kept in pairs or in groups. There are advantages and disadvantages to horses being on their own in paddocks or with other horses but the advantages of keeping horses together outweigh the disadvantages. Horses can injure each other if they are paddocked together – but this is usually in the initial stages and in most cases they settle down after a while. When horses are paddocked separately, they will attempt to get close to or even play with one another by running up and down the fence line. They will also try to groom each other over the fence, which can also be dangerous. Fences can and often do cause deadly injuries to horses.

By putting horses in together rather than having each horse in its own paddock also means that very small paddocks can be amalgamated, making it easier to manage and improve pasture. The group of horses can then be rotated around the larger paddocks. If you have limited pasture it is better to have horses in yards for part of each day. They can be turned out to graze together for the other part of the day. The grazing time can be increased when the availability of grass is good and decreased when it is not (see Chapter 6). Having yards (see p. 115) enables you to manage your pasture much better.

Paddocked horses should be visited at least once a day to check for changes in body condition, as well as signs of ill health or injury. This check enables you to spot small problems before they develop into larger ones. The water supply must also be checked at least once daily, more often in hot weather.

Figure 2.11 Horses grazing together Credit: Jane Myers

Stabled/yarded horses

On a small property it is usually necessary to confine horses for part of the time in order to avoid land degradation and to maximise the grazing potential of the property. Unless there are lots of horses and very little land you should be able to manage your horses so that they spend varying amounts of time in paddocks and yards/stables. Over time, with correct property management, the available grazing should increase allowing the horses more time to graze.

There are many issues that you need to be aware of when confining horses to either stables and/or yards that will have a large impact on the quality of life and the health of your horses. Yarded horses have the advantage over stabled horses of having access to more fresh air; otherwise the issues are similar for both situations.

The issues involved with confined horses arise because confinement is an unnatural situation for a horse. Horses are designed to move around a large area and forage for a large part of each day. Confining horses can lead to problems as set out below.

- Circulation problems ('filled legs') due to a lack of activity. In the natural state a horse walks for many kilometres per day which keeps the circulation moving.
- Behavioural problems due to lack of activity and reduced eating time and lack of companionship. Natural living horses, as well as being free to move and forage, live in herds and have a rich social life. Boredom or stress is common as a result of not being able to exhibit normal behaviour (see p. 10).
- Gastrointestinal problems such as colic are common if the horse is not allowed to eat a large amount of forage. Many confined horses spend long periods of time between 'meals' when in fact a horse should have access to forage on an 'ad lib' basis (see p. 21).
- Thrush is a common problem in confined horses. Stabled/yarded horses need more regular foot care (cleaning) than paddocked horses. This is due to their being unable to avoid standing in manure and urine (see p. 28)
- Respiratory problems due to airborne pollutants are a problem in confined horses (see p. 20)

In order to offset some of the problems of confinement, horses should be turned out regularly (for part of each day) into a paddock or if this is not possible an exercise yard such as a round yard or a safely fenced arena. In fact turning an over-exuberant horse out into a large yard before turning out into the paddock will reduce land degradation as the horse will then get straight down to the task of grazing rather than galloping around. Even if the horse is being worked every day try to allow the horse some turnout time. If you have no turnout, for whatever reason, consider agisting the horse until you do.

Stabled horses should be able to see and preferably touch another horse. Modern stables usually prevent this from being possible. Design and build stables where horses can interact and groom one another over the wall. If space permits, have yards attached to stables so that horses can go outside into the fresh air and interact with each other over the fence.

A confined horse needs attention at least twice a day. Early in the morning and as late as possible in the evening in order to spread out the visits and feeds. If possible the horse should be turned out for either the daytime or overnight. In cold wet weather it is best to turn out in the daytime. In hot weather it may be better to turn the horse out overnight because it is then cooler and there are fewer insects after dark.

On each visit the horse will need to have the water checked and cleaned/topped up if necessary, feed topped up and the stable or yard cleaned. Always remove a horse from the stable when doing these chores to reduce the amount of airborne pollutants that the horse is exposed to.

Airborne pollutants

Ensure that individual boxes are large and have good ventilation. Airborne pollutants in stables are a combination of gases, bacterial and fungal toxins and spores, mites, pollen, feed particles and animal components. These come from sources such as urine, manure, feed (hay and other feeds), bedding and animal skin, respiration, and outside dust blowing in (bringing with it soil and manure). In addition, stabled horses may lie down longer due to the extra time available which leads to an increase in intake of airborne pollutants (the air quality is worse nearer to the ground). Horses are highly sensitive to airborne pollutants and tend to be exposed to them for much longer than other domestic animals.

Attempt to keep airborne pollutants to a minimum. Dust and fungal spores originate mainly from bedding and hay. Poor quality hay is dustier and has more fungal spores than good quality hay. The hay can be soaked or steamed just prior to feeding to reduce fungal spore inhalation.

Bedding is another area that can add to airborne pollutants. The main types of bedding for stables are straw (wheat, barley, oat), hard or soft-wood shavings, or sawdust. Other less common alternatives are shredded paper and commercial beddings such as cleaned, dust-free shavings or compressed wood pellets which tend to be more

Figure 2.12 Horse-friendly stables, with good roof ventilation and open windows. Horses can interact over the walls and the higher corner protects them while feeding Credit: Annie Minton

expensive. Straw is usually dustier and contains more fungal spores than wood shavings and sawdust and is not as absorbent. Shredded paper has also been used with some success being absorbent and not dusty. Commercially produced beddings have the lowest dust rating but are more expensive. If they are used in conjunction with rubber matting/flooring, however, they can work out to be economically viable as very little is needed in this case (see p. 140).

Feeding and watering

This section looks at the practical aspects of feeding and watering horses and will also look briefly at what to feed. See Chapter 12 for more detailed recommended reading on the subject of feeding.

You need to be able to assess the body condition of your horse by eye and this is something that when practised becomes automatic whenever you look at your horse. Always be alert for signs that your horse is 'dropping condition' or getting too fat. If you rate your horses every day using condition scoring – a system that enables you to check the body condition of your horse just by looking at it (see Table 2.1 and Figure 2.13) – you will be able to determine when the horse needs extra feed or a reduction in feed. Ideally your horse should have a body condition score of around 3 – good.

Feeding

Horses that are underweight or overweight are at risk of serious health problems. Correct feeding is just one of the aspects of taking care of your horse's health but it is a very important one. Feeding horses correctly does not have to be too complicated. Horses need a very high roughage diet (for example hay, grass or silage) and only require high-energy feeds such as grains and processed feeds (concentrates) when they are working hard and growing. Aim to feed the best quality roughage that the horse can eat without becoming overweight. In addition your horse should have access to salt and minerals on an ad lib basis.

Paddocks can sometimes look as if they contain lots of feed when they don't. For example the first rains after a dry period can turn paddocks a nice shade of green which leads some owners into thinking the horses have plenty to eat. This short, new, soft feed has little food value. Another misconception arises when looking at 'horse sick' paddocks which appear to contain lots of grass. The long grass in the 'roughs' is unpalatable and is unavailable for grazing because of the grazing behaviour of the horse (see p. 6)

As the weather in autumn slowly starts to get colder, horses can lose weight alarmingly if they are not managed carefully. More good quality hay is usually necessary to help horses maintain body condition in cold weather. Teeth care is important and rugs may be necessary for some older horses and horses that are not carrying good body condition already.

Horses that are overweight are not healthy animals either. In addition to wasting money, feeding a horse too much can lead to unruly and dangerous behaviour due to having too much energy. Horses that are overweight also have a higher risk of developing laminitis (see p. 36).

Table 2.1 Body condition scoring system

	Neck	Back and ribs	Pelvis and rump
0 Very poor	Marked ewe neck	Skin tight over ribs	Angular pelvis – skin tight
	Narrow and slack at base	Spinour processes sharp and easily seen	Deep cavity under tail and either side of croup
1 Poor	Ewe neck	Ribs easily visible	Rump sunken, but skin supple
	Narrow and slack at base	Skin sunken either side of backbone	Pelvis and croup well defined
		Spinour processes well defined	Deep depression under tail
2 Moderate	Narrow but firm	Ribs just visible	Rump flat either side of backbone
		Backbone well covered	Croup well defined, some fat
		Spinous processes felt	Slight cavity under tail
3 Good	No crest (except stallions)	Ribs just covered	Covered by fat and rounded
	Firm neck	No gutter along back	No gutter
		Spinous processes covered but can be felt	Pelvis easily felt
4 Fat	Slight crest	Ribs well covered – need firm pressure to feel	Gutter to root of tail
		Gutter along backbone	Pelvis covered by soft fat – felt only with firm pressure
5 Very fat	Marked crest	Ribs buried – cannot be felt	Deep gutter to root of tail
	Very wide and firm	Deep gutter	Skin is distended
	Folds of fat	Back broad and flat	Pelvis buried – cannot be felt

(Source: Horse Sense, 2nd Edition)

Some useful tips for feeding are shown below.

- If feeding concentrates, use large flatter containers (tyre feeders are good). This works especially well for greedy horses as it slows them down (they are not able to grab huge mouthfuls of feed as they can in a narrow deep bucket).
- Feeders should have rounded corners for easy cleaning and horse safety, and should be deep enough to prevent feed being nosed out and wasted.
- Feeders should be positioned so that you do not have to walk back past a horse that is eating once you have put the feed into the yard or stable.
- Feeding horses before turning out on pasture will lead to them taking longer to start eating. Not feeding first will lead to them getting down to grazing quicker, which is more desirable.
- If feeding on sand, feed above ground height or on large rubber mats to reduce sand ingestion which can lead to sand colic.
- A horse needs approximately 1.5 to 3 per cent of its bodyweight in feed (dry matter) per day.
- A horse should chew for at least 12 hours per day to keep behavioural and health problems at bay.

(a) 0 = very poor.

*Very sunken rump; *Deep cavity under tail; *Skin tight over bones; *Very prominent backbone and pelvis; *Marked ewe neck

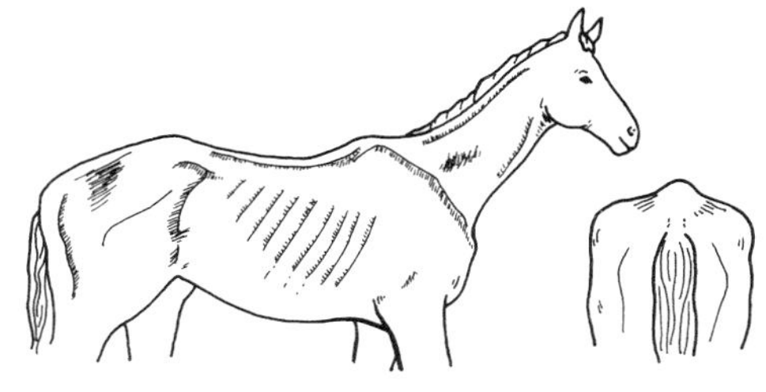

(b) 1 = poor.

*Sunken rump; *Cavity under tail; *Ribs easily visible; *Prominent backbone and croup; *Ewe neck, narrow and slack

(c) 2 = moderate.

*Flat rump either side of backbone; *Ribs just visible; *Narrow but firm neck; *Backbone well covered

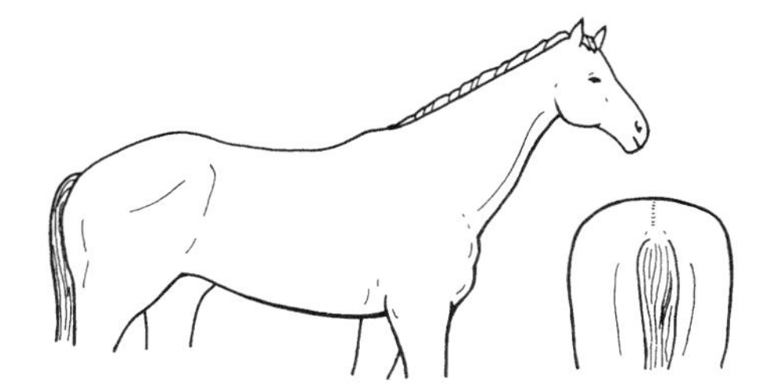

(d) 3 = good.

*Rounded rump; *Ribs just covered but easily felt; *No crest, firm neck

(e) 4 = fat.

*Well-rounded rump; *Gutter along back; *Ribs and pelvis hard to feel; *Slight crest

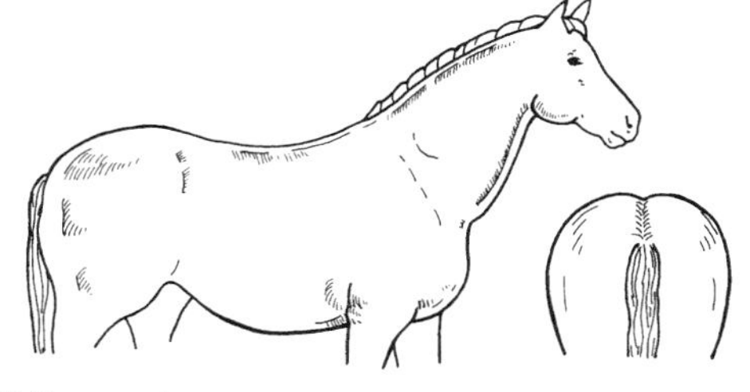

(f) 5 = very fat.

*Very bulging rump; *Deep gutter along back; *Ribs buried; *Marked crest; *Folds and lumps of fat

Figure 2.13 Body condition scoring system

Source: *Horse Sense*, 2nd Edition

- Horses are susceptible to poisoning from contaminated and mouldy feedstuffs. Horses cannot vomit so any ingested feed must go through the whole system before being eliminated.

When horses are given supplementary feed it is better to feed them in individual stables or yards. This is because it reduces the incidence of fighting as horses get much more competitive over supplementary feed than grass. Dominant horses will threaten and cause injury to subordinate horses at feed time, particularly if they force them into a fence. Feeding horses individually means that they will be able to be provided for according to their needs rather than their wants (i.e. the biggest bully gets the most feed)!

If supplementary feed must be provided to groups of horses in a paddock situation, feed the horses in individual feeders spread wide apart. The safest way to provide feed to a group of horses is to have the feeders attached to the fence in a line so that feed can be put in the feeders without anyone having to enter the paddock with feed.

The diet of a horse should be made up of a least 70 per cent roughage (grass, hay or silage). A horse that is not working very hard can live on a 100 per cent roughage diet. A horse that is working hard can be supplemented with concentrates (grains and processed feeds) that comprise up to 30% of the diet. Ideally a stabled or yarded horse should have roughage available all of the time. Pasture hay is better than lucerne in this respect because it is lower in energy, therefore the horse has to eat more of it which occupies the horse for longer.

If stabled/yarded horses are receiving concentrates (in addition to their roughage), they should be fed at the same time as each other at regular intervals. If possible, concentrates should be split up into at least three meals a day, with the last meal being as late at night as possible. This is also a good opportunity to check that the horse has enough clean water and roughage to last the night.

If feeding hay on sand or sandy soil, you may need to feed it off the ground in some sort of feeder such as a rack, haynet or large container to reduce the chances of sand colic (see p. 36). If hay nets are used to feed horses they must be tied high enough so that they do not hang low when empty enabling horses to get their foot caught in them. Large tractor tyres make excellent feeders for hay. If there is a risk of sand ingestion the bottom can be lined with a tarp or rubber sheet.

When feeding hay to a group of horses the hay will need to be spread out so that all of them can access it. This may mean having several piles of hay or feeders.

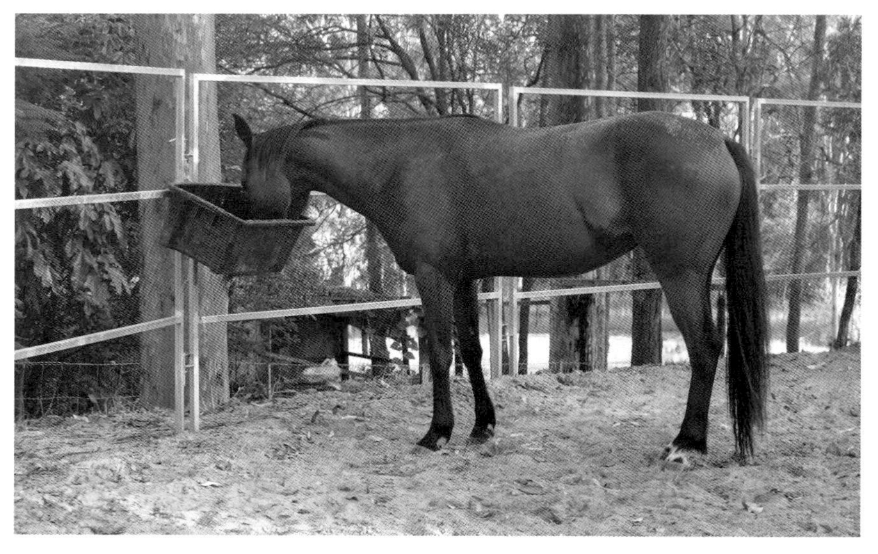

Figure 2.14 Feeding on sand — Credit: Jane Myers

Figure 2.15 Hay feeder Credit: The Horse Shed Shop

When choosing your method of feeding, keep in mind that a horse should spend as much time as possible with its head in the grazing position. This is because this posture allows the airways to clear naturally. Low feeding ensures drainage of the lungs. Horses (like many large grazing animals) have long necks and are prone to problems if they eat for long periods with their head at body height.

Only buy the best hay from a reputable source, as hay that contains weeds will infest your paddock rapidly. Even if fed in yards or stables, some weed seeds will end up in the paddock via the droppings. The money saved by buying cheap hay will go nowhere on the costs of renovating your paddock (and all the ones adjoining).

Unless your property can produce enough extra fodder to make hay or silage (see p.105) you will need to buy forage in. Good hay should smell sweet and have some green colour. There should be no evidence of rotting and mould. Good hay will feed your horses well and in addition add nutrients to your property (via their composted manure) and may add grass seeds as well depending on your feeding method. Poor hay will lead to poor horse health and may introduce new weeds to your property. Some poisonous weeds are more palatable to the horse when dried (in hay) such as ragwort and fireweed. When buying hay, check for weed content and ask what steps the producer has taken to prevent weed infestation on their land.

Figure 2.16 Green hay Credit: Jane Myers

Feeding silage (which is sometimes called other names such as haylage) to horses in Australia is still in its infancy compared to other countries (such as New Zealand and in most of Europe) due to there being some unfortunate incidences with botulism many years ago.

Figure 2.17 Silage Credit: Jane Myers

Silage is ensiled fodder or grass. It is commonly fed to cows and can be a very good quality feed for horses. When fodder is ensiled rather than made into hay it is baled much sooner after cutting, while it still has a high moisture content. It is then wrapped in airtight plastic sacks (usually as round bales) that preserve the fodder and, as long as the bag is not opened, it can stay as fresh as the day it is made for many years. Once opened a bale must be consumed within a few days so unless you have a few horses a large round bale will spoil before being eaten. As silage becomes more popular in Australia it may become available in smaller bags that are packaged especially for the horse industry, as is the practice in New Zealand and Europe.

Botulism occurs when soil or rodents are taken in with the silage at baling time. The fodder must be cut high enough to reduce the chances of this happening. Botulism thrives in the anaerobic environment of a bag of silage. Vaccinating any horse that is being fed silage can prevent death by botulism. Although the botulism bacteria cannot be seen, discarding any sour-smelling, burnt-looking or slimy silage is also recommended to avoid possible digestive disturbances in the horse.

The advantages of silage over hay are that it:

- is dust free.
- has a more consistent quality due to not being as reliant on the weather
- is a high quality fresh feed in which the nutrients are preserved, and
- does not require a building to store it and keep it in top quality.

Figure 2.18 Drinking Credit: Jane Myers

Watering

All horses must have access to clean drinking water 24 hours a day (see Chapter 5). Lack of water will cause dehydration, it can also cause impacted colic (due to drying of food in the gut and slower digestion), therefore an adequate water supply, in terms of both quantity and quality, must be provided to meet your horses' needs.

In confined areas, water should be situated where it will stay as clean as possible. This is usually a distance away

Figure 2.19 Exercising a horse Credit: Jane Myers

from the feed. The airborne pollutants created when horses are confined, and especially when stables or yards are cleaned out, taint water so it must be changed frequently irrespective of whether you have automatic waterers/drinkers or you are using buckets. Make sure the horse cannot tip buckets over by placing them in a tyre.

Horses have a very good sense of smell and taste and will refuse to drink, even to the point of dehydration, if their water supply is polluted, stagnant or even if the water supply changes suddenly. Observe horses that are new to the property to check that they are drinking.

Exercise

The amount of exercise that you will need to give your horse depends on whether the horse lives in a small area or a large area. Horses that are confined to stables or yards for much of the time must be exercised daily. This exercise can take various forms but the most important thing is that the horse moves and maintains its fitness. Horses that live in paddocks do not necessarily have to be exercised every day if the paddock is large enough for the horse to exercise itself by walking around and by having the occasional run.

Teeth care

The horse's diet, mainly tough fibrous material, requires a lot of chewing and grinding. This causes the teeth to wear down and often creates problems such as sharp edges and uneven wear. If the horse's ability to grind down food sufficiently is compromised for any reason, the enzymes and microbes of the gastrointestinal tract have a hard time continuing the digestive process and the result is a drop in condition.

Teeth problems can also cause behavioural problems as the horse attempts to alleviate any pain. Horses need regular dental care if they are to get the maximum benefit from their feed and perform well.

Potential problems include those shown below.

- Sharp cheek teeth (molars) – this occurs to some extent in all horses but its occurrence is accelerated when horses have a high grain diet.
- Imperfect meeting of the teeth (parrot mouth) – this is where the front teeth (incisors) do not meet. This causes a problem with grazing and the horse often develops sharp 'hooks' on the molars due to them also being out of alignment.
- Wolf teeth – teeth that are a much-reduced vestige of a tooth that was well developed in the ancestor of the horse. They sit in front of the first molar and because they usually have shallow roots they can be loose. A loose wolf tooth may cause a horse to head toss or be reluctant to respond to the bit.

- Teething problems – as with human babies, the eruption of teeth in young horses may cause transitory trouble. The horse may have 'caps' which are temporary (milk) teeth that have not fallen out but form a cap on top of the permanent teeth.
- Decayed teeth – this can lead to the destruction of the tooth, which may lead to infection of the surrounding bone.

Some of the signs of dental problems include weight loss, loss of coat shine, irregular chewing pattern, quidding (dropping balls of food out of the mouth while chewing), unresponsiveness to the bit or head tossing, excessive salivation, bad breath, swelling of the face or jaw, lack of desire to eat hard food and reluctance to drink cold water.

A vet or qualified horse dentist should be called in regularly to thoroughly examine and carry out any necessary work on the teeth. The teeth of a mature horse should be checked at least once every 12 months if it is grazing only. Horses under the age of five and horses that are fed large amounts of concentrates should have their teeth checked every six months.

Hoof care

Care of the horse's feet is one of the most important aspects of horse care because a lame horse cannot be used, hence the old saying 'no foot, no horse'.

The hooves are able to adapt to various surface and climatic conditions. For example, hard feet result from hard dry ground conditions.

Any signs of lameness should be investigated and treated promptly. A minor injury may cause the horse pain and it should be given a rest from work until the lameness disappears. A veterinarian will need to be called for most situations where the horse is lame unless you are experienced at dealing with certain conditions. Make a habit of regularly checking the horse's legs and feet as this provides a chance to notice any abnormality such as swelling or heat. Never neglect wounds, punctures or significant cracks.

Some of the signs that a horse is lame include:

- 'nodding' the head, especially in trot, as the horse attempts to keep the weight off the affected limb
- resting a front leg by 'pointing' it forward or bending it by resting the toe on the ground (it is normal to rest a hind leg but not a front leg)
- uneven gait
- not willing to move
- sudden severe lameness, which can mean a hoof abscess.

Lameness can be caused by many factors such as laminitis, abscesses in the hoof, deep cracks, thrush (a yeast infection of the frog) and injuries to the legs. Some conditions of the hooves do not necessarily cause lameness straightaway, such as thrush and 'seedy toe', which is a cavity at the toe due to separation of the laminae. These conditions need attention or they will lead to lameness. As a horse owner you need to learn about these conditions so that you can recognise them.

The feet of a stabled or yarded horse require cleaning (picking out) more often than if the horse lives in a larger area (such as a paddock). This is because in a small area the horse is forced to stand in manure and urine-soaked bedding (no matter how often you clean the stable or yard) and this compacts in the feet and can cause problems such as thrush. A horse that is kept in a large grassy paddock will either have clean feet or will develop packed soil in the hooves depending on the conditions in the paddock. Both are a perfectly natural occurrence and it is usually only necessary to pick out the feet just before you ride.

Oiling the feet is not usually necessary as it can actually prevent the feet from absorbing moisture (the hoof absorbs moisture from the outside). Oil will not cause the hooves to grow better, as is often proclaimed by manufacturers. If you must oil the feet on a regular basis it may be necessary to soak the feet in water first so that they do not dry out. It is actually more damaging for the hooves to be constantly switching from wet to dry conditions than to either be continuously wet or continuously dry.

To shoe or not to shoe

Not all horses need to be shod but they all need good foot care, so you will need to use a farrier even if you do not intend to shoe your horse. Some horse's feet can be conditioned to working on hard and even stony ground if time is taken to condition them and the feet are trimmed properly. Other factors to bear in mind are that unshod hooves do less damage to paddocks than shod hooves, horses wearing back shoes should not be turned out together, and horses wearing shoes are more likely to injure themselves, for example a shod hoof can catch on fencing wire causing serious injury to the hoof.

Unshod horses require attention from a farrier either every six weeks or longer if the hooves are wearing down at the correct rate. It all depends on how abrasive the surface is that the horse lives/works on and the hardness of particular horses' hooves.

If a horse is shod it is very important that the shoes are taken off and replaced or reset after no more than six weeks. The hoof is unable to wear down with a shoe in place, therefore the hoof gets broader and longer. This causes many problems for the horse including separation of the laminae, potential damage to the bones inside the feet and the tendons and ligaments of the legs.

A good alternative to shoeing is to use hoof boots. These are slowly gaining acceptance, even with the owners of hardworking horses and they can also be very useful for a horse that is only used periodically but needs protection when working. If these boots are to be used they must fit well as they can cause rubbing if they are too large or small. See Chapter 12 for details of hoof care websites.

Rugging

There are pros and cons associated with rugging horses. Rugging is necessary in some cases, i.e with horses that get 'the itch', for some older horses in inclement weather, extreme cold and so on, and they can provide some protection against biting insects. However, many horses would be far more comfortable if they did not have to wear rugs. Some of the factors to take into consideration include your local climate, your budget, the work the horse does, the sensitivity of the horse to biting insects, where the horse lives (in/out/both) and how much time you have available.

Many horses are over-rugged and get overheated. If you feel hot, then your horse certainly will. It is as simple as that, even more so as they already have a coat to keep them warm.

Only rug a horse if you have to. Your horse may not look as smart or clean as you would like but it is natural for a horse to be muddy or dusty. If you do rug your horse, make sure that any rugs fit well and be aware that a rugged horse will actually require grooming more often (see below).

Grooming

Grooming involves brushing the whole of the body in the direction of the hair growth to remove mud and dust, picking out the feet and tidying the mane and tail with a brush. Brushing the mane and tail too often and too vigorously will lead to too much hair loss.

The amount of grooming that a horse needs depends on its living conditions. A horse that is constantly rugged needs to be groomed more than a horse that is not rugged. Irrespective of whether the horse lives mainly in a stable or in a paddock, the rugged horse must be groomed frequently. Rugs prevent the dead skin that is shed daily from being sloughed off. Horses roll and groom each other to shed dead skin. This mutual grooming also provides massage. Rugs prevent the skin from being shed when rolling and the horse from receiving the benefits of massage, therefore this service must be provided by the owner. It is a good idea to allow rugged horses to periodically (preferably every day) spend time without rugs to have a good roll and get dirty. This can be done when the horse is sweaty after work. The horse can then be cleaned up and rugged again. A good tradition in racing stables is to let the horse roll in a sand roll after exercise. The horse is then hosed and rugged again.

Unrugged horses should be cooled down and dried with a towel after exercise, if sweaty, before being put away. After exercise on a warm day, the sweat can be removed with a sponge or a hose (no detergent) and the horse returned to the paddock to roll in the dust.

Trimming the horse will make it look smarter but you must be careful, as some of the hair you regard as scruffy is essential to the horse. You should never trim the inside of horse's ears (even though many people do) as this hair guards against insects, dirt and rain. Because the ears of the horse point upwards they need the protection of hair (even if the horse is mainly stabled, as there is a lot of dust in the stabled environment). You can neaten the appearance of the ear by trimming the hair that protrudes when the ear is held closed.

The whiskers around the eyes and muzzle should *never* be removed (again many people do this) as they are essential as feelers for the horse.

Parasite protection

The main parasites that horses can be afflicted by are ticks, lice, parasitic worms and bots.

The three main *ticks* that can affect horses are cattle ticks, New Zealand cattle ticks (bush) and paralysis ticks. Cattle ticks are found in Queensland, north-east NSW, the Northern Territory and Western Australia. The New Zealand cattle tick (also called the bush tick) occurs in Australia and New Zealand. In Australia it is found in the coastal region of south-east Queensland, the NSW coastline and north-eastern Victoria.

Paralysis ticks are usually found in coastal areas of eastern Australia in late winter, spring and summer. There have been occasional outbreaks in New Zealand. These ticks cause paralysis and can kill a dog or even a young foal but adult horses have some resistance.

Some affected horses will rub the area excessively. Horses in tick-infested areas must be checked regularly and if any ticks are found apply a pyrethrum-based insect repellent. Wait about an hour and then apply the solution again. The tick should drop off soon after, or if not you can then remove it with tweezers as it should be dead. Do not pull ticks off when they are still alive as this may result in the head being left under the skin. If the horse shows any adverse effects then a veterinarian should be consulted. If you live in a tick-free area you must report the tick finding to your local agriculture department.

There are certain restrictions on moving horses in Queensland between tick-free and tick-infested areas (see Chapter 12 for website details about moving horses in and out of tick-infested areas of Queensland).

Lice can affect any horse but those that are low in condition and unhealthy are more susceptible. Lice can be seen around the mane and tail area and the affected horse may rub itself. Treatment involves powders or washes that can be obtained from feed or saddlery stores and it is usually necessary to treat all of the horses on the property even if only some show signs. Rugs and saddle blankets will also need treatment.

The major parasitic *worms* that affect horses are the large and small strongyles, tapeworms, roundworms and pinworms. Some of the common signs of parasitic worm infestation are tail rubbing, pale gums, ill thrift, colic and poor coat. Worms can also cause diarrhoea or sudden death.

Parasitic worms rely on a specific host to complete their life cycle. The cycle of most equine parasitic worms involves eggs being passed out in the dung which then hatch out as larvae, the larvae attach themselves to grass that is then ingested by the horse. Once in the horse the larvae migrate through the organs of the body (different worms have different behaviours) causing damage as they go. Eventually the larvae, after becoming adults, end up in the digestive system of the horse and deposit eggs that pass out with the dung. Thus the cycle goes on.

Controlling parasitic worms is best approached from several angles. Some of the things that you can do to minimise the worm burden of your horses are:

- worm your horse regularly using an effective product. It is better to use a broad spectrum (and usually more expensive) product less often than a cheaper narrow spectrum product more frequently.
- collect droppings daily, or every two days at the most, in very small paddocks (see p. 190).
- avoid feeding on the ground if there is manure present.
- harrow paddocks in the hot dry months and/or when the temperature is very low to break up droppings and kill worm larvae (see p. 191).
- rotate horses to fresh pastures or to pastures grazed by cattle or sheep (see p. 112).
- do not overstock horse paddocks (see p. 83).

Bots are often thought of as worms when in fact they are the larval stage of the bot fly. Bots are not as harmful as worms but they still require regular control.

Bot flies are active in summer, laying eggs on the lower body of the horse, particularly on the legs, shoulders and belly. The eggs are the size of a pinhead and yellow in colour. They can be seen clearly on an affected horse.

When the horse bites itself or mutual grooms with another horse that has them, the eggs are taken into the mouth. After burrowing in the gums for some weeks, the larvae are swallowed and attach themselves to the stomach wall for several months before being passed out in the dung to develop into adult flies. The egg-laying activities of the adult flies cause considerable annoyance to horses in summer although they do not sting the horse. Although bots cause mild ulceration at their attachment sites in the stomach and on very rare occasions these ulcers may perforate, infestations generally present no problem to most horses.

Although a tedious task, it is possible to remove bot eggs from the horse by scraping or clipping, but this should be done in a yard, and not where horses graze. Horses should be treated for bots twice a year, therefore use a boticide in early winter when fly activity has ceased and the larvae are in the stomach and again in the spring. Boticides are included in certain parasitic worming products. Check the product that you are planning to use to see if it includes one.

Pest protection

Flies and other insects

There are many different types of flies that can make your horse's life a misery if they are not controlled. They can cause horses to stamp, twitch, kick, lose weight, bite at themselves, roll excessively and they may even panic and run. They can cause eye infections and will invade any open wounds and sores on a horse. They are also a health hazard to humans.

Figure 2.20 Fly traps Credit: Jane Myers

An excessive number of flies are usually due to poor manure management. They may be breeding at local chicken farms or on your property.

Horses need protection from flying/biting insects, this can be in the form of mesh veils, mesh ear covers, repellents, fly sheets and so on, depending on how large the problem is. Mesh veils must be removed at night as they reduce the night vision of a horse. Horses should not be turned out without another horse or some form of cover if biting flies are present. When horses are turned out together they can stand nose to tail and protect each other's faces with their tail.

Other flying insects that affect horses are mosquitoes and midges that bite and cause itching. Some horses have an allergy to midge bites (called sandflies in some areas) and this condition is called 'The Itch' or Queensland Itch.

Other things that you can do to reduce insects are cover manure, ensure there is no stagnant water, and dig out and remove wet soil areas, i.e. where horses urinate. Another helpful strategy is to grow herbs and flowers that repel pests (see p. 181). Homemade repellents can be made from diluted citronella oil and other solutions such as cider vinegar.

Products can be purchased to help control flies and other insects (see Chapter 12). Electronic devices can be fastened near stables and yards that attract and kill flying insects (some cover up to one acre). These devices kill insects either by electric shock or drowning. Shade cloth can be hung across stable doors to reduce the number of insects entering the stables.

Encourage wildlife that will eat flying insects by creating habitat for them. Frogs eat large numbers of flying pests so make sure you place rocks for shelter near waterways. Encourage birds and bats which also eat large numbers of insects (see p. 174).

Other pests

Rodents can be a problem where horse feed is stored. However, by keeping the area very clean these pests will not be as common. Rodents can be controlled in many ways including poison. Poison must be used very carefully as it can and often is inadvertently picked up by other animals. Place it in pipes so that pets cannot get to it. More humane and organic forms of rodent control involve using other animals such as terrier dogs or cats (however, cats have many drawbacks as they tend to kill all small animals indiscriminately). In some parts of Australia a resident native python is very handy as it will eat and control rodents without being dangerous to humans.

Inoculations

Horses can be vaccinated against two infectious diseases that are potentially very serious, namely tetanus and strangles.

Tetanus (sometimes called lockjaw) is usually fatal and is caused by a bacterium that lives in soil and body tissues. Horses are very susceptible to tetanus. This bacterium is anaerobic which means it lives without oxygen so certain types of wounds, such as deep punctures, are more prone to the development of tetanus. Vaccination injections (Equivac-T) are given as two injections, four weeks apart followed by a third a year later. Then boosters are given every five years after that. Initial immunity occurs three weeks after the first injection. Horses can start a vaccination schedule from three months of age.

Strangles is a contagious disease that is more prevalent in foals and young horses. Affected horses have discharges of pus from the nose and develop abscesses under the jaw that eventually burst and discharge pus also. Many horses recover but some have chronic illness and even die. Vaccination (Equivac-S) requires three injections two weeks apart, followed by yearly boosters. The vaccination is not an absolute preventative, but in the case of an outbreak vaccinated horses are either not affected or have a milder form of the disease and recover more quickly. There is a two-in-one vaccine available (Equivac 2 in 1) to make it easier to vaccinate your horse against both tetanus and strangles at the same time (see Chapter 12 for details of a website about vaccinations).

Assessing health

If you suspect that your horse is ill you need to be able to carry out a basic clinical examination. These findings can be discussed with your veterinarian over the phone and can be used to help determine the need for a visit. A basic clinical examination involves assessing the horse's vital signs, and other indicators such as the coat, nasal discharge, eyes, manure and urine, and behaviour of the horse. To appreciate changes, you need to know what is normal for your horse, so you should examine it when it is problem-free to be ready for the day when problems do occur.

Prevention of ill health is better than cure, so it is important to:

- check your horse as often as possible
- ensure that stables and paddocks are safe
- keep records of your horse's past injuries etc. and have a good routine of feeding, watering, exercise, worming, vaccinations (tetanus and strangles), hoof care and dental care
- know your horse's normal vital signs (body temperature, heart rate, respiration rate, hydration status, mucous membrane colour and gut sounds), so that if you suspect your horse is ill you can compare what is normal for that horse with your findings
- constantly evaluate the appetite, body condition, coat, nasal discharge, faeces, urine and behaviour of your horse
- regularly check for lumps, bumps, swellings, abrasions, cuts, sores and wounds, sore spots, hair loss and itchy areas.

Coat

The coat of a healthy horse lies close to the body and shines (some colours of horses shine more than others, for example bays and blacks shine more than greys and roans). If the coat is staring (standing up) then this usually means that the horse is ill and/or cold. The coat hair stands up in an attempt to trap air and warm the horse. In the short term the condition of a horse's coat is not really affected by illness, however, in the long term it is. A horse that has been ill for a long time and/or has very little body condition will have a very poor, dry-looking coat. Diet can also affect the coat.

A horse should not be sweating unless there is a reason for it, for example if it is a hot day and/or the horse has just worked.

Nasal discharge

It is quite normal for a horse to have a trickle of liquid in the nostrils. However, this should be clear and not sticky. In the unhealthy horse there may be yellow or green sticky mucus pouring from the nose.

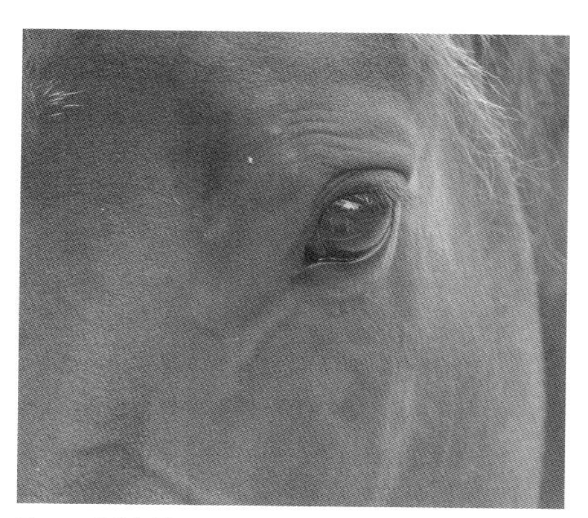
Figure 2.21 Clear eyes Credit: Jane Myers

Eyes

The eyes should be clear, not running or sticky with discharge, and alert when awake.

Manure and urine

A healthy horse will pass manure 8–12 times a day. The colour will vary depending on what the horse is eating. Urine varies from pale yellow to darker yellow and can be clear or cloudy. Urine and manure should be passed without straining or signs of discomfort. Mares urinate more often when in season.

Behaviour

The behaviour of individual horses varies enormously, so again it is important that you know what is normal for each particular horse. A healthy horse should be alert, interested in what is going on around it, sociable with other horses and forward going when being ridden. Other signs of health are a good appetite, lying down to sleep in the sun and interacting with other horses.

When a horse is ill it will not display its normal behaviour. A sick horse will stand for long periods with its head hanging down. The horse may lie down more than normal or not at all. The sick horse looks 'tucked up'; this is where the flanks appear to be sucked in. The horse will either not be alert or may be excessively anxious, it may keep looking at its flanks and/or paw the ground (signs of abdominal pain). Other behavioural signs of ill health include excessive chewing on objects and sudden aggressive behaviour. Basically any changes in normal behaviour should be investigated.

Common health problems

Colic

Colic is a general term meaning abdominal pain in the horse. Colic is often a secondary condition to some other problem, for example a horse with laminitis can develop colic and the reverse is true also. Colic can either be mild or very severe and it is important that you recognise the signs so that a vet can be called if the case is anything other than mild. Even mild cases can progress to severe later on. When a case is severe the bowel has either twisted or may twist very soon. This condition either requires immediate surgery or destruction of the horse for humanitarian reasons.

Horses with colic are usually restless, pawing the ground, stamping, looking around and kicking at the belly. They also may frequently lie down and get back up again; they may even sit like a dog. Other signs include refusing to eat and sweating due to the pain. The horse will have an increased heart rate and this can be a guide to the severity of colic. In acute cases the mucous membranes of the gums become darker and congested. There is usually an absence of droppings but in some cases the horse passes more droppings than normal.

Colic can be caused by many situations such as stress, eating poisonous food, blockages in the gut from fibrous food that has not been chewed properly due to poor teeth, sudden changes of feed, irregular feeding, overfeeding, worms or sand in the gut (sand colic). Horses develop sand colic if they are fed on sand or graze sandy pastures and take in significant amounts of sand as a result.

A common cause of colic is when horses are fed a diet that is too low in roughage. The gut cannot cope with periods without roughage passing though it and colic is often the result. Only feed high energy grains if the horse is working at a level that calls for such feed. Even if feeding a high energy diet, always try to keep the percentage of roughage as high as possible. One way in which this can be done is to feed as much grass hay as possible (rather than more energy dense hays such as lucerne). As the grass hay is lower in nutrients the horse has to eat more of it. This means that the horse spends more time eating/chewing and is better off both from a behavioural as well as physiological perspective.

If signs of colic appear and persist for longer than a half an hour, call a vet. It will help if you can asses the vital signs before doing so. Allow the horse to lie down if that it what it wants to do as this is often the most comfortable position. However, try to prevent the horse from rolling excessively as this can lead to complications.

Laminitis

Laminitis (sometimes called founder) usually affects both front feet. Sometimes just one front foot is affected in cases where one front leg has had to take more weight due to the other front leg being injured. It is possible for laminitis to affect the hind feet without affecting the front feet though this is very rare.

In a severe case, the horse stands with the forelimbs stretched out in front of the chest and hind limbs under the abdomen. This stance (the 'sawhorse' posture) is an attempt to relieve the weight on the front feet. There will be heat in the feet and a bounding pulse can be felt at the main artery to the foot at the level of the fetlock. The horse will be reluctant to move and may lie down.

When laminitis occurs there is a breakdown in the bond between the sensitive and insensitive laminae in the foot and this leads to rotation of the pedal bone which can even penetrate the sole.

Conditions predisposing to laminitis include:

- grazing unlimited amounts of lush pasture
- gorging on starch-rich grains
- obesity and lack of exercise
- travel stress
- intestinal disorders resulting in acute diarrhoea
- retained afterbirth in recently foaled mares
- surgery
- non-weight-bearing lameness in one leg
- the use of some drugs.

Laminitis is a complicated condition and every horse can get it. The condition can leave your horse permanently debilitated or the horse may need to be euthanased as a result of it. A horse owner should learn as much as they can about this condition (see Chapter 12 for recommended reading on laminitis). A vet should be called immediately if you suspect that your horse has laminitis.

Prevention is vital as this is such a serious disease. During spring you *must* keep a close watch on how fat your horse is becoming – green lush grass is high in soluble carbohydrates and is very dangerous in unlimited amounts. You should aim to maintain your horse in good, but not fat, body condition (see p. 21). The crest should not be allowed to get too large and hard and you should always be able to feel the ribs. If the horse or pony is getting too fat, take preventative measures before it reaches the critical point. You will need to reduce the feed intake of the horse by placing it in a paddock with less feed or even a yard (with low energy hay) for a period of the day (see p. 111). Exercise is also important in reducing the horse's weight, as weight reduction will be a very slow process unless the horse is worked.

When horses are being fed large amounts of grain you should always reduce the feed intake when the workload drops off. The risk of laminitis will be minimised if the roughage (older, drier grass or hay) content of the horse's diet is kept high. Watch out for signs of stiffness and hoof soreness that may indicate the early stages of laminitis.

Wounds

All wounds should be cleaned as soon as possible with cold running water. The cut or torn skin should be placed back where it was before being injured, and held in place with a dressing, cotton wool and bandage. If a wound is to be stitched it needs to be done shortly after the wound occurs and certainly within four hours. Call the vet immediately.

Respiratory infections

If your horse has a runny nose or cough, stop exercising or working the horse. If the horse is off its feed or the cough/runny nose symptoms persist for more than a few days, you need to get a vet to look at the horse. Working a horse that has a respiratory infection can lead to chronic coughing and permanent damage to the respiratory system if ignored. Possible problems can include coughs and colds, virus infections, strangles, pneumonia and heaves/chronic bronchitis. Pneumonia can lead to permanent damage to the lungs or even death.

Eye problems

Eye problems are very painful for the horse, and can quickly become serious, so require prompt attention. Signs of problems include partial eyelid closure or changes in the appearance of the surface of the eyeball. Healthy eyes are essential for all working horses. The safety of the horse and of the rider depends upon good vision.

Some of the problems that horses are susceptible to include ulcers, uveitis, conjunctivitis, injuries to the eyeball or eyelid and blocked tear ducts. Any changes to the appearance of the eye should be investigated thoroughly and owners should not put off seeking veterinary attention.

Sprains and bruises

It is important to control any swelling and inflammation for the first few days after an injury that results in sprains or bruises. Cold water or, preferably, ice packs are very good in this regard. The water or ice packs should be applied for a minimum of 20 minutes at least three times a day. At other times the affected area should be kept bandaged if possible. As well as reducing swelling and inflammation, ice is very effective in minimising pain.

Some rest from work will usually be required to allow time for the injury to heal. A vet is the best person for advice in this regard.

Skin conditions

Horses can suffer from various skin conditions. You need to check your horse's skin on a regular basis for signs that a problem is developing. A healthy horse has a gleam to the coat even if dusty or muddy. Signs that indicate skin problems include dry flaky skin and coat, itching, dermatitis, lumps or growths on the skin, hair loss and weeping sores, swollen areas and blisters. A vet should be called for any skin conditions that you are unsure about and are unable to deal with yourself.

See Chapter 12 for recommended reading and helpful websites on horse care.

3

Property selection

Buying a property on which you can keep your horses is a dream come true for many people. However, a horse property can be a sponge for money so it is important that you chose the right one – careful selection in the first place can save a lot of heartache and expense.

Try not to rush into your decision, although this can be difficult if the market is moving fast. Aim to view several properties, the more properties that you see the better you will get at spotting potential problems, and when you do find your ideal property you will be able to make an informed decision.

In order to avoid the perils and pitfalls consider getting help – this may be in the form of an experienced friend (someone who has owned/set up a horse property before) or an expert, i.e. a consultant. If possible, talk to people who already have a horse property. Ask them if they have any tips and whether there is anything they wish they had known before they bought. You need to bear in mind that different people value different things, what constitutes a good property depends on people's personal preferences and viewpoint.

The following points are the sort of things that you need to either think about or find out about an area or a particular property before taking the next step and starting the buying process. Some of these points, such as zoning or future proposals for the area, will be rechecked by your solicitor. However, if you do your homework first you will be able to reject properties at the early stages if they are unsuitable without incurring costs from a solicitor until you find the right property.

This chapter looks at property with a view to buying. Certain subjects, such as fencing for example, are only touched on here because they are covered in much more detail further on in the book. It would be a good idea to read the whole book before buying a horse property so that you understand the work and costs involved.

Finding a property

Where can you find properties for sale? Agents' offices are a good place to start. Properties are also advertised in newspapers, on the internet and in magazines (some horse magazines have a property section).

Properties can either be listed with an agent or as a private sale by the owner. Private sales are becoming popular as they can have advantages for both buyers and sellers. Because there are no agents' fees involved, the price of the property may be slightly lower or more negotiable than a property listed with an agent. Another advantage is that owners of properties (especially acreage) can usually tell you a lot more about a particular property than an agent can. Even if you are using an agent, it is sometimes a good idea to talk to the owners (if this can be arranged through the agent), as there is a lot more to an acreage property than simply a house on a block. The agent will only be able to tell you so much.

Many people still prefer to use the services of an agent. The agent acts as a go-between for a buyer and a seller. Agents range enormously in the level of service that they give. As most agents are paid by the seller, and are therefore working on behalf of the seller, it is worth considering getting a buyer's agent who will look after your interests. This is happening more frequently, especially when the buyer is looking for land with particular attributes.

Some agents specialise in rural properties and can be a wealth of information about a given area. On the other hand some agents, particularly those on the urban fringe, do not usually have a good understanding of the issues involved with acreage properties (i.e. they only know about houses not land) and may encourage you to look at properties that are far from ideal. This can be frustrating. However, over time you will build a picture of certain areas and their suitability and also certain agents and how to read between the lines.

When you are first starting to look it is a good idea to drive around an area that you may be interested in, this gives you a feel for the locality. While you are there you can call on local agents and start to build a picture of what is available. Knowing the area helps when agents contact you about properties. You will already have an idea about whether you will consider that area and it can save you a lot of time.

Figure 3.1 For sale Credit: Jane Myers

Local people are an invaluable resource when you are looking for a property. Shop owners, and in particular the local horse feed store, are good places to start. Local people can give you a lot of information that the agents might not.

Look out for open days on properties in the district that you are interested in. This is a good way of comparing properties and starting to build a picture of what an area is like. Even if you do not buy the property it is an opportunity to gather ideas on designs if it is already set up as a horse property.

Make up a wish list in order of importance of things that are of interest to you, for example proximity to good riding areas or proximity to a pony club. It is unlikely that you will find a property that has everything on your wish list (unless you have an unlimited budget), but as long as it has the features that are most important to you, you may decide to compromise on the rest and either make improvements later on or use the property as a stepping stone to another one in the future.

The sort of property that you end up with will depend on many factors including how much money you have available, the availability of local tradespeople, your own skills and abilities, as well as your willingness to have a go when it comes to improving the property.

Even if you are planning on living in the property indefinitely, make sure that you buy something that would be reasonably easy to sell on, as you never know when you may have to change your plans.

Your budget

Your budget will ultimately decide which property you can buy. If your budget is tight make sure you are not buying a property that has lots of features that you do not really need, for example eight stables when you only plan to have two horses. The more features the property has, the more it will cost. Also if you are borrowing to buy the property you will be paying interest on features that you are not using.

When looking at a property you need to bear in mind questions such as: *Is it good value for money? What is the amount needed to be borrowed? How much will it cost to service the loan?* Remember that work may be harder to get in the new area that you are planning on moving to. Also keep in mind that some country areas are more difficult to get finance for. You may need a larger deposit than you would with a suburban property. Shop around the lending institutions, particularly ones that have offices in rural areas (these tend to be more helpful about rural properties), as there is a huge variation on who will lend what.

When you find a possible property to purchase you need to evaluate the price of the property and take into account all of the features that will save or cost you money in the future. Compare the price of this property with others in the area that have sold recently.

Developed or undeveloped land?

Is it best to go for an already established 'horse property' or a block of land that has not been used for horse-keeping previously? There are advantages and disadvantages to both scenarios.

Established 'horse properties' are understandably more expensive than properties without facilities. Buying an established property can save you lots of time in the future. However, unless the facilities are well designed and built with good quality materials you may be paying an over the top price. Properties that have previously been used for horses may already have problems such as compacted soil, weeds and so on depending on the importance that the previous owner attached to land care. You may also find that you do not like the layout of the facilities once you start to use them but changing them will be expensive (plus you will then have paid twice). It is generally easier to build from scratch than to alter old stables or yards, if they were poorly constructed in the first place.

You need to weigh up whether you are prepared to pay the higher cost of buying a property that already has horse facilities and will be immediately usable, against buying a property that has no facilities which will mean that you have to build them (see Chapter 7).

Some properties are already well fenced but not how you would like them, for example timber, white painted fences are nice to look at but are very high maintenance. Fencing generally has a limited life span so make sure you are not paying for fences that have reached the end of their productive life. If all of the fences are in a bad state it may cost you several thousand dollars to re-fence the property. You need to budget for this.

Fences should separate wet paddocks from dry paddocks (changes in land type), and creeks, rivers and dams should be fenced off from the paddocks (see Chapter 8).

Properties that have not previously been used for horses will of course have no facilities. However, this means you will be able to design the property and build facilities to your specifications. A particular problem with land that has not been developed for horses is that it is often fenced for cattle with barbed wire. In this case you will need to either re-fence it before moving your horses in or use temporary electric fencing until you are able to re-fence.

If it is an undeveloped block, check that the block has a potential house site. You may need to get professional help to advise you. Some of the factors that help to determine a house site include slope, microclimate, water supply, soil type (stability), drainage, fire risk and views.

If the block is subdivided from a farm, it is possible that the farmer may have subdivided the most troublesome part of the larger property. It is therefore wise to check for waterlogged soil and other land degradation problems (see p. 90).

Another potential scenario is a block that already has a house on it but no horse facilities. This can be a good option, provided that the house is positioned well on the block – you would have somewhere to live while you got straight on with developing the rest of the property.

Figure 3.2 A horse property Credit: Jane Myers

Figure 3.3 A selection of horse properties

Credit: Jane Myers

Property location and size

There is a link between location and size of properties because land becomes more valuable the closer it is to a town. A common decision is whether to go for a larger property further out or a smaller property nearer a city or regional town.

If you plan to stay working in the same job, you need to decide how far away from your place/s of work you can realistically live. It is no good buying your dream property then spending lots of precious time driving to and from work. Also other family members may not be too enthusiastic about being so far out of town, unless they are also keen horse lovers.

Before buying you need to be sure about what size of property you need. This will help you to find the right one. For example if you buy a property that is too large it may become a burden. However, a much more common scenario is to buy a property that is too small. Land varies enormously in its carrying capacity (see p. 83) but it is easy to underestimate how much land you will need. This may mean having to move again sooner than you planned, or having to compromise on the way that you keep your horses or the number of horses that you keep. As a guide an acre equals 0.405 hectares – a hectare is very close to 2.5 acres.

Remember that the smaller the property, the more space the house and buildings take up relative to the total size. The house, garden, sheds and stables can easily use up one to two acres. Therefore a five-acre property will usually have only three to four acres available for pasture, a three-acre property may only have one to two acres. If an arena is added then that is even less space available for pasture.

Be realistic about how many horses you may end up with and remember that each paddock must have several periods of rest each year if you want to maximise your grazing. Ideally you should allow for some cross-grazing as well (see p. 112).

As blocks of land decline in size, the relative price per acre/hectare increases dramatically so a bigger property does not always cost much more. However, many subdivisions are five or ten acres and often this is all that is available for sale within a reasonable distance of a city or regional town.

Figure 3.4 Keeping horses in the suburbs Credit: Jane Myers

Many horse properties can be found in suburban areas. Why would anyone want to keep horses in suburbia? Keeping horses in the suburbs can have its advantages and disadvantages. With careful planning and management, horses can be kept successfully in this environment.

Some of the advantages of suburbia are the proximity to services such as shops and so on, and it may mean not having to drive as far to work. Non-horsey members of the family are able to live without the disadvantages that country living offers for people that have no interest in the countryside. Some of the disadvantages of keeping horses in suburbia include the expense of suburban land (such properties tend to be very small, i.e. less than two or three acres) and the proximity to neighbours.

The suburban property usually involves a more intensive system of management than a property that is not surrounded by other houses. It must also be kept spotless if the horse owner is not to raise the ire of neighbours.

Individual property features

Natural features

Most natural features are a bonus. A property that has a running creek, trees, and a lake will have shelter and water already in place and will be a more attractive property for having them. Trees and shrubs are essential on a property and a property with none will take some time to grow.

These natural features should be protected (by fencing off) before you allow horses to have access to the property. Horses can cause extensive damage in a short space of time to certain natural features (see p. 66)

Trees that are spread out on a property are harder work to protect than trees that are grouped and can be more easily fenced (see Chapter 9).

Large rocks on a property are not as much a problem as small rocks that will make it difficult to use machinery in the paddocks. Rocks have a habit of working their way up to the surface continuously, so even when an area has been cleared rocks may still keep popping back up.

Figure 3.5 Natural features of a property Credit: Jane Myers

If the property is timbered you may not be able to clear it for paddocks. Quite rightly there are more and more restrictions on clearing timber and even when the area is cleared you will still have to establish pasture, which can take some time. Pockets of natural vegetation on a property that is otherwise cleared are a bonus as they provide privacy for the property owners, habitat for wildlife, act as shelterbelts and enhance the appearance of the property. It is far better to improve the already cleared land and leave the pockets of native vegetation (and protect them with fencing) than to clear them for extra grazing.

Any natural ecosystems on the property should be regarded as a huge plus when you are looking to buy. If you are planning to make your property sustainable then the presence of already established natural ecosystems means that you only have to protect them rather than create them in the first place. As you will see from the rest of this book, these ecosystems will enhance the way in which you keep your horses and your lifestyle; therefore their protection should be regarded as imperative on a property.

The landscape

Undulations make a property far more interesting aesthetically and some undulation is desirable in a horse property. A balance of flat and sloping land is best – it makes a property more attractive and having hills means that there will be drier areas in the wet months on hilltops. Undulations are also beneficial to horses as they aid fitness. The presence of steep hills, however, is not desirable, as this will lead to erosion. Horses are not designed to stand and graze in steep country all of the time, also sloping land creates more run-off. A slope of 2–6 per cent is ideal, a lesser slope will tend to be boggy and a greater slope will lead to erosion.

Look at which way the hills slope (the aspect), in the Southern Hemisphere north-facing slopes face the sun, giving warmth and good vegetative growth.

Any excavations (such as for the house and an arena) are more expensive to carry out on a slope than on flatter land. However, a property that has a combination of steep and not so steep land can be planned so that the house is built on the steep area (a pole home is a good option in this case). This will give good views and the less steep land (2–6 per cent slope) can be developed for horse keeping, including an arena (if you are planning to build one).

Figure 3.6 Hilly land

Credit: Jane Myers

Figure 3.7 A bore Credit: Jane Myers

An undulating property will have better views. A good view can greatly increase the value of a property. However, for many horse people the best view of all is to be able to see your horses from the house. This can also save time, as you are then able to do quick checks of horses without having to go outside every time. Situations when this is very useful are for example when you have turned a new horse out, or when a mare has foaled. Being able to observe your horses easily is one of the best aspects of owning your own horse property.

Water supply

Water is a very important commodity. On a country property it is essential you are self-sufficient for water. Otherwise, when it runs out you will have to buy it in. This is expensive if you are watering stock in addition to your domestic usage.

When considering a property, first of all find out how it gets its water. There are many possible sources and if a property has more than one it is more likely to be drought tolerant.

Town (mains) water (if the property is near a town) is for domestic purposes. In some areas it can be trickle fed, where it is pumped to the property at a slow rate and this water is stored in a tank. The same tank can also collect rainwater. More common is full pressure mains water.

Tank water means that the water run-off from the buildings is collected in one or several tanks. This can then be used for stock and domestic purposes.

Bore water varies in quality and it can only be used as domestic drinking water if it has been tested and passed as safe. Some properties rely on bore water, either for watering stock or sometimes as drinking water for humans as well. If the property does not have a bore, find out if neighbouring properties have them. If so, it is likely that you would also be able to install one (you would have to check with the local authority first). You may not need it, but it is good to know that it is an option if necessary.

Dams are for watering stock and can be used for some domestic purposes such as flushing toilets. Ideally dams should be fenced off and the water pumped to holding tanks from where the water is distributed. Check any dams for signs of leakage.

If the property borders a creek or river the property may or may not have a licence to pump water from it. Find out if it has a licence and how much water can be taken. Just because there is a creek on the property it doesn't mean that it will run all year. If the creek is described as a 'seasonal creek' this means that it will dry up in the drier months.

Check out any irrigation (ask to see it working), if the property is set up for it. The water used for irrigation is usually supplied by a bore, dam, creek or a river.

If a property does not appear to have enough water then you need to calculate the costs involved to obtain more, i.e. buying more storage tanks, sinking a bore or putting in more dams. You can then decide if the property is still going to be suitable, given the extra expense.

Check that there is access for water tankers to deliver water if necessary, such as in a drought situation. Even though you are hoping to be self-sufficient for water you need to be prepared for the worst (see Chapter 5).

Pest problems

Speak to locals about pests such as ticks and flies and anything else the area might be prone to. For example properties that are situated near chicken/egg farms can be inundated with a type of biting fly that is often called a 'stable fly' but actually breeds most prolifically in chicken manure. Most areas have some kind of pest but some have a particular problem with a particular pest, especially if an ecosystem is disturbed resulting in certain pests becoming dominant (see p. 32).

Natural environmental disasters

Before buying your land, it is a good idea to find out what the risks of natural environmental disaster are for that district. If you already own a property and you do not already know these things, you need to find out. Many areas are at risk of fire, flood, drought or violent storms, and the design of a property should account for the probability of their occurrence and the impact that they will make.

Some areas are very prone to flood. Find out if the district you are interested in has flooded before. You will not be able to build on a flood-prone area even though many older houses come up for sale in those places. The council will be able to advise you on where you can and cannot build so that you are safe from floods. See Chapter 12 for useful websites on protecting your property and horses from fire and flood.

Property layout

No two properties are the same and when you go to view properties you will see some interesting layouts. Sometimes there is a good reason for the layout of a property that at first sight appears to be unusual. The facilities may have (or should have) been placed to take into account wet areas, microclimates and so on. However, some properties are just badly designed due to being built on an ad hoc basis without much thought being given to the layout. It is difficult to determine these factors in a short visit and this is one area where you can learn a lot by speaking to the owner. In order to gain more insight, ask if they built the facilities themselves and why they chose that particular layout.

Make sure that the property has good all weather access. Any steep dips on the property (such as over creeks) may be impossible for longer vehicles such as a car and float to access. They may also be impassable in wet weather. You will also need to think about access for heavy vehicles such as hay carters. If the property access is difficult it is a good idea to get quotes to see how much it will cost to make the crossing usable in all weathers.

Beware of buying a property where the only access has poor visibility when turning out onto the road. A vehicle pulling out of a property with a float is much slower than one without, so the entrance must allow for this (see p. 61).

If the property is situated on an unmade road find out who owns it and who is responsible for its maintenance. Usually this is the local council and the road is main-

Figure 3.8 Paths and bare areas Credit: Jane Myers

tained by them. However, you need to know if this is not the case and what would be expected of you in this scenario.

Pastures and soil

Be aware that many properties come on the market at the best time of the year, such as when the new grass starts to grow in the spring, because that is when they look their best. The same properties may be very dry a few months later. Alternatively properties that have drainage problems (waterlogged soil) may be presented for sale in drier times. Even a paddock full of weeds can look great if it is mowed. You must inspect the pastures carefully.

Find out what the land has been used for in the past. If the property has been a farm there may be old sheep or cattle dips on the land which are very dangerous and must be removed. The property may have been used for intensive cropping and therefore undesirable chemicals may have been used on it. If you find out anything like this you need to contact and get advice from the agriculture department for that area before going any further with a possible purchase.

Figure 3.9 Bare soil Credit: Jane Myers

Land that has been disturbed (i.e. ploughed) is more likely to have weed infestations. From a sustainable point of view, the best land is land that has not been used for intensive farming or, if it has, it has been managed without or with minimal use of chemicals. This kind of land has usually previously been used for cattle (beef, low intensity) and as long as it has not been overstocked in the past should be reasonably easy to manage and improve.

The quality of the pastures and soil will affect factors such as surface drainage, the amount of feed and weeds. Good soil is a huge bonus. It will support pasture, grow trees and bushes, require less maintenance than poor soil and will allow you to maintain and further improve your pasture using such strategies as cross-grazing. Soil can be improved but it does take time.

Find out if the soil is clay, loam, sand or gravel, with sandy loam being preferred. Check the land for signs of waterlogging (signs of pugging i.e. where animals' hooves have left deep holes when the land was wet). One indication of waterlogged soil is the presence of plants that survive in wet soil, such as reeds and rushes (see p. 91).

Ask if the soil has been tested recently and consider having some tests done (if time allows), as this will help you later on with planning your property. Soil tests tell you the pH of the soil and which nutrients are present and which are lacking. They will also show up poor drainage (waterlogging) and salinity. Visual checks for salinity include looking for scalds or bare patches, or even white crystals shining on the surface (in a bad case). Valleys of cleared land are prone to salinity. Other signs are dead and dying trees, declining pasture (clovers and even weeds dying) and there may be lots of barley grass. Ask a local landcare group or other expert for advice if you suspect salinity (see p. 95).

Most properties have weeds of one sort or another. When a property is presented for sale it may have been slashed so that weeds are not evident at a glance. Walk into the paddocks to see if it is grass or weeds that have been slashed. Slashed weeds will tend to reveal a lot of stems with spaces between them, whereas a grassy paddock will usually still look grassy even when slashed. Also remember that many weeds are only present at certain times of the year so you may not be able to see them at the time of inspection. Ask what has been done to prevent/control weeds. Look across at neighbouring properties for evidence of weeds.

Go to see the local agriculture department. Find out what the particular weed problems are for the area, especially declared/noxious weeds. Find out how extensive they are, what threat they pose to the land and animals, how to eradicate them and how easy or difficult this will be. Find out who has the responsibility for eradicating different classes of weeds. In some areas you may also be responsible for controlling the weeds on the roadside outside the property.

If the property appears to have a large proportion of problem weeds rather than pasture, think carefully about buying, as you will spend a lot of time and expense trying to eradicate them. Weeds are also indicators of other problems such as soil compaction and unbalanced soil nutrients (see p. 99).

Asking the previous owner about how they manage the property will tell you a lot about how much land degradation there may be. If the property has been used for many horses and good management practices such as paddock rotation were not carried out then it is very likely that there will be some land degradation. Even good land cannot cope with continuous pressure from grazing animals. If the previous owner outlines what steps they were taking to manage the property this tells you that they are aware and responsible.

Poor management would show up as bare compacted areas in the paddocks and waterways, signs of erosion such as surface erosion and tunnel erosion (land that is prone to tunnel erosion is not suitable for horses), uneven growth in the paddocks with areas of roughs and lawns and a large number of weeds. Another telltale sign is unthrifty horses – however, some properties can be in very bad shape and the horses still look wonderful because they are fed on lots of supplementary feed (see p. 90).

Try not to buy someone else's problems unless you have time and money to spare. On the other hand, if the property is a good price, turning it into a good sustainable horse property will be an extremely rewarding experience – but if this is to be your first horse property you may want to start with something a little less challenging.

The wider area

Climate

Even though the climate for a particular region is reasonably predictable there can be variations (micro-climates) within it, so you need to find out both the macro and micro-climate of your property. For example some areas are windier than others and some tend to miss out on rain on a regular basis. To find out such things speak to the locals if possible.

Some of the climatic factors that are useful to know about an area include:

- the annual rainfall
- the amount of frost
- the number of chill hours
- the length of the growing season
- the temperature extremes
- the altitude (higher is colder)
- prevailing wind direction and conditions.

Access to riding areas

Many properties are sold as 'horse properties', yet they are situated on a fast busy road. If you are not planning to ride off the property at all this may not be a problem, but if you plan to trail ride finding the right situation is very important. In this case try to find a property that at least has access to quiet country roads (not all country roads are quiet) or better still, access to State forest/forestry land. Properties with access to public forest are very much in demand so they can be difficult to find and tend to sell quickly. If a property backs onto a national park be aware that you may not be able to ride there as many national parks have restricted access. Find out what government department controls this area and contact them. You can then find out what the access will be to this area (if any) and if there are any plans in the pipeline to change its status.

Neighbours

Find out what the neighbours do with their land. For example they may have a stallion that is kept in a paddock backing onto your potential property. If the neighbours have horses you will probably need to erect a double fence with a bushy screen between them. On the other hand they may be trail bike riding enthusiasts which can lead to problems with noise.

Figure 3.10 A riding trail — Credit: Jane Myers

Is the next-door property messy? This might not be a problem in the short term, however, it may make your potential property difficult to resell in the future, especially if the mess creeps nearer to your property over the years. If the property next door has not yet been built on, try to find out where they are planning to build. An amazing number of people build right next to the fence line on acreage which will mean that your property has reduced privacy (privacy being one of the benefits of living in the country).

Find out about the surrounding area. You need to know all the pluses and minuses of the district including any noxious industries (these can be pig farms, feedlots, slaughter houses and chicken/egg farms). Unless these properties are practicing sustainable management systems there may be an enormous amount of flies in the area and bad odours, especially on the days that sheds are cleaned out. Crop and fruit farms can also be a problem if they spray harmful chemicals (unless they are organic, this is usually the case).

Utilities and services

If you are looking at buying a block of undeveloped land that is not part of a new subdivision, check the availability of power (unless you are planning to use solar) and phone. Just because power is on the block next door does not necessarily mean that it can simply be extended to your potential property, extra junction boxes or power lines may have to be added which can be expensive. Check with the power company. New subdivisions usually have power and phone to the gate.

If there are no properties with amenities nearby then it may be very expensive to get them connected. You will need to contact the electricity and phone companies for quotes and base your decision on that.

Ask what the local council rates are for the property and what they cover. For example it is common on country properties to have to transport your own rubbish to the tip. Find out if there are recycling facilities available.

If the property has a septic tank for sewerage find out where it is placed. Sometimes the entrance to the septic tank inadvertently gets covered over when a garden is landscaped. This is a problem when the septic tank needs to be cleaned out.

Other things to find out are the whereabouts of facilities such as shops, hospitals, doctors and schools and if there is any transport to them. For example does a school bus run come near the property? Find out if the property has mail delivered or whether you would have to open a box for mail (at a post office or elsewhere).

Zoning and restrictions

Contact the local council and find out what the zoning is for your potential property's area. Then find out how this will affect you, for example if it is zoned water catchment there will be restrictions about where you can build and what you can do with the land.

Zoning controls what is built where so that an area develops in a regular way (or stays undeveloped as the case may be). Zoning establishes certain areas for certain uses such as industrial, rural, rural residential etc., and also protects areas such as water catchments. Zoning will dictate factors such as building type and how far built structures can be placed from a road or creek. You will need to get council approval before building anything and zoning is one of the regulations that council uses to grant or withhold approval. If you build without council approval you can be made to take the building down (at your expense). Find out if there are any plans to change the zoning in the future.

Ask to see a copy of the land title to find out if there are any covenants or easements on the title deeds. The agent or the owner, if selling privately, will have a copy of these.

Find out from the council if there are any other restrictions that will affect the property such as the number of horses that can be kept on the land, restrictions on land use, local covenants, local government restrictions on the number of dwellings on the land and property development, such as recycling gray water and composting toilets. The area may be heritage listed thereby restricting renovation and development.

Future proposals for the area

Find out if there are any development proposals in the pipeline, such as road widening/urban expansion, wind farms etc. Future development may be a plus or a minus depending on your viewpoint. Either way, you need to know.

Find out if the potential property or its neighbouring properties can be subdivided in the future. Again this can be a plus or a minus. At present the property may have just one neighbour on each side however if that land can be subdivided the property will then have many neighbours and many potential problems.

Find out if there any easements planned for the area that may run across or very near the potential property such as gas and power lines.

Legal issues

Buying the property

If you decide to go ahead and buy the property after you are happy with the initial checks, you will need to contact a solicitor or conveyancer to do the legal paperwork. They will do thorough check-ups through local and State planning departments. They should also check that the buildings have the necessary permits. They will check for any outstanding rates or builders' orders on the property.

Have the property inspected by buildings and pest inspectors (agents may be able to give you some contact details, although they are not allowed to wholesale recommend any particular one). Termites (white ants) are a problem in parts of Australia and New Zealand. It is well worth paying to have the property checked out thoroughly before you buy and discover that you own a white elephant.

Insurance

It is important that as soon as you sign a contract on the property you have it insured. In the event that anything happens to the property in this period you may be classed as the owner (even though you have not yet paid for it). Insuring an established horse property is more expensive than just a house on a block as there is a lot more to insure. Just before you move on to the property with your animals, you should also obtain insurance that covers you for third party insurance for accidents caused by your animals straying into the road or other people's properties. Rural property insurers usually offer this as an insurance package i.e. home/contents, facilities/fences and third party for any damage your animals cause.

Settlement period

Try to arrange the settlement period to suit you and include any conditions on the contract that you think may be necessary. For example the usual settlement period is about six to eight weeks. It is possible to lengthen this period and sometimes to shorten it, if it suits both parties.

You may want to stipulate certain conditions, such as any rubbish/weeds to be cleared before settlement, paddocks to be left in a certain condition and so on. These things are negotiable between the buyer and the seller and should be established at the negotiation stage before signing a contract to buy.

Surveying

Before settlement you should walk the boundaries of the property and check out the location of the survey posts. There should be a white post dug into the ground at each corner and a marker for any changes of direction to the shape of the block. If the property was surveyed recently there will also be white sticks with plastic ribbons on them at intervals between the corner pegs. These tend to disappear over time. The corner posts may be overgrown if the property has not been surveyed for some time. It is your responsibility to satisfy yourself that the survey posts correspond to the perimeter fence and that the block is the size stated. Your solicitor/conveyancer should ask if you have done this.

Moving in

Now all you have to do is move in. If you currently keep your horses on agistment or at livery, it is a good idea to arrange for them to remain there at least for a few days or a week after you have moved in so that you can complete the move without having to supervise them at the same time. Try to arrange that they arrive no later than midday so that they have plenty of daylight hours to get used to their new surroundings. Either way it is a good idea to keep them confined in yards or stables for the first night in order to minimise potential accidents.

4

Property design and management

Designing your own horse property can be very rewarding, but it can also be frustrating because if you make the wrong decisions you have to live with them. The more time spent in planning and preparation will lead to more informed decision-making. So take your time and gather as much information as possible. You won't regret it.

Correct location of fences, laneways, buildings and yards is essential for the environment as well as the efficient operation and management of a horse property. If it is well planned, it will be safe and efficient and will save stress and money through reduced labour and running costs. It will also be more aesthetically pleasing, a much better place to live and therefore a more valuable piece of real estate.

It may be helpful to refer to books on permaculture design when designing your property. Permaculture design can be applied to a horse property with benefits to all parties. Basically permaculture is about creating harmony between landscapes and people, increasing sustainability and reducing waste (including wasted effort). This fascinating subject cannot be covered in detail here however see Chapter 12 for where to go for more information.

When planning your property, remember it may take many years to get it where you want it. Enjoy the journey and see it as a never-ending road, at the same time be flexible and accept that situations and your own ideas will change.

Influencing factors

When planning a property there are numerous factors that you should take into consideration. If you check that each decision you make takes these factors into account it will help you to decide more wisely. These factors can be loosely grouped into the following:

- **Environmental factors**: How the environment will affect the property and vice versa. Planning to have a minimum negative impact on the environment and, where necessary, to enhance the environment should be a major concern. The elements (wind, water, earth and fire) and how to manage them come under this heading as do the management of flora and fauna. Examples include where vegetation for shade needs to be planted, what surfaces to put where to reduce dust and mud, how to reduce the fire risk.

- **Economic factors**: Your budget will determine what you can do and when you can do it. Planning will help you to separate projects into order of importance. For example planning to build an area that the horses can be confined to (yards with shelters for example) as soon as possible will at least reduce the damage that horses can do if allowed free range access to a small property. Another example would be using electric fences at least as a temporary measure, as they are both economical and effective. This measure will free up money that would otherwise be spent on permanent fences and instead you can begin to plant trees which should be a priority.
- **Ergonomics**: The organisation of your working environment for ease of use, for example the logistics of moving horses and possibly vehicles around the property. Provision needs to be made for feeding out, handling horses and parking a float/s. It takes into account where the house, buildings and other horse facilities will be situated in relation to each other in order to reduce extra work. Areas that will be used most frequently need to be identified. Ultimately good ergonomics leads to a better lifestyle as good design reduces hard work.
- **Safety factors**: This means planning to build facilities that will be safe for you and your horses. To achieve this you need to take other factors such as horse behaviour into account. For example if paddocks come to a narrow point at the entrance there is a much greater risk that horses will get kicked by another horse (this situation is also very dangerous for handlers) when leading them in and out of the area.
- **Aesthetics**: what a property looks like is important. It is a much nicer experience to live and work in a beautiful environment than an ugly one. Planning with the environment in mind will lead to an aesthetically pleasing property as they go hand in hand, for example a block with no trees, polluted water and bare overgrazed paddocks is ugly whereas trees, clean water and green paddocks are beautiful.

Making a plan

Whether you are designing from scratch, or improving an existing property, planning is the first crucial step in the process. It is too easy to rush off and buy trees and have fences put in only to realise that you have done things back to front, it is better to minimise wasted time and money by doing things right the first time.

Make a plan before starting any major changes or additions. It should be accurate and detailed. The best starting point for a good property plan is an aerial photograph, such as those produced by one of the private companies that now provide this service.

Using this photograph and tracing the shape of the block and its features onto a sheet of paper, a series of plans can be made at various stages of property development. If you are unable to obtain an aerial photo then sketch the shape of the block onto a sheet of paper, including the property features, as shown below.

- Landscape features such as valleys, slopes, gullies, ridges, contour lines, trees and shelterbelts and water (dams, creeks, and rivers).
- The climate, for example prevailing winds and the angle of the sun in summer and winter.
- Anything on the outside of the property boundaries that could affect the property such as neighbours (and their proximity), a dirt road (dusty and noisy), an intensive farming operation (possible chemicals, smells from sheds etc.).
- Different land classes (even small properties usually have different types of land especially if the property is undulating), for example rocky, swampy, heavy clay and sandy areas.
- Problem areas such as erosion (tunnel, water or wind erosion) waterlogged soil, compacted soil (areas where animals tend to congregate), habitat for vermin (e.g. rabbit warrens) and weeds.
- Fixtures such as buildings, existing horse facilities, power lines, easements, bores, fences and drainage lines.
- Any zoning requirements that will affect where you can or cannot build, for example part of the property may be classed as water catchment.

Make up plenty of copies of this basic plan so that you can try out lots of different designs.

When developing your property plan, seek all the advice available. Once you have your completed plan aim to show it to people who have some experience in designing and/or owning a horse property and get their opinion. Try to visit as many horse properties as possible to get ideas. Contact the relevant government agencies for advice or your local Landcare group. Also courses can be undertaken at TAFE/Technical Colleges on 'Whole farm planning' to assist in this process.

A small property can have some or all of the following features.

- Areas for holding horses i.e. stables/shelters/yards
- Areas for grooming, tacking, clipping and shoeing
- Roads and tracks in/out and around the property i.e. laneways/driveways
- Areas to graze horses i.e. paddocks with pasture
- Areas for exercising and training (arenas/round yards etc.)
- A source and storage of water (dams/creek/river/bore/tanks etc.)
- Areas for tack, feed and equipment storage
- Areas for humans i.e. a home/gardens
- Land set aside for wildlife/shelterbelts/windbreaks/firebreaks
- Areas for other domestic animals such as dogs.

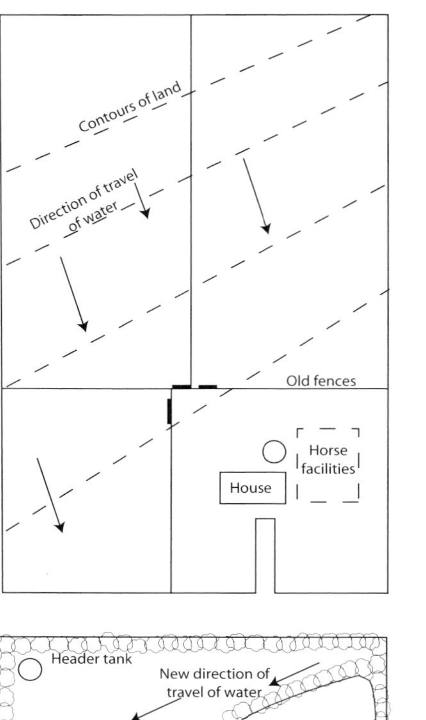

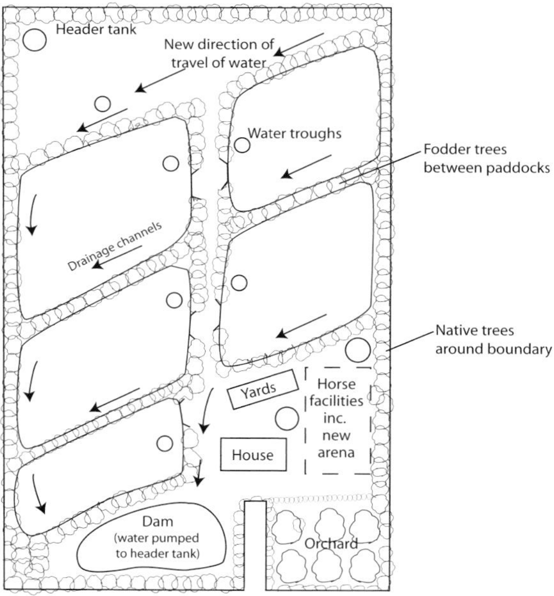

Figure 4.1 A property plan. Credit: Jane Myers

Top: A 15-acre property with sloping land. There are waterlogging problems around the house and horse facilities, due to water travelling downhill towards them. Fences are in the wrong location.

Bottom: The same property after pasture renovation and revegetation, the addition of new trees, dam, arena, covered yards and an additional water tank to the buildings and a header tank at the highest point of the block. Water is channelled towards the dam. More paddocks allow for greater rotation and better pasture management. New fences follow the contour lines. The house is surrounded by the dam, arena and yards, and an orchard, and this, in conjunction with fire-resistant trees, increases fire protection.

Fire planning

All properties, particularly those in Australia, are at risk from bushfire on a regular basis. Plan to make the property as fire resistant as possible. This involves the correct placement of buildings and shelterbelts as well as regular routine maintenance.

Fire can either come to the property in the form of a bushfire, or it may originate from the property in the case of stable, house or barn fires. Fire is often something to which not enough thought is given until it is too late. Good property design and carefully worked out plans for what to do if it occurs are essential if you are to minimise the impact of fire.

Some of the things that you need to find out (from neighbours, CFA/RFA, Bureau of Meteorology, Landcare office) are given below.

- When is fire likely to occur in your area?
- How often does it tend to occur?
- Is there a warning period?
- Does it follow a predictable pattern?
- Which direction do the hot dry winds tend to come from? Fire usually comes from a specific direction. However, you must also be prepared in the event that it may come from a different area.
- Is the property part of a community fire-reporting network?

If you are planning a property from the beginning you will be able to implement fire safe features as you build. An established property may require some alterations to make it safer both for fire prevention and in the event of a fire.

Think of your property in terms of 'zones'. The inner zone includes any dwellings, stables, yards and barns (the most valuable parts of the property) surrounded by a buffer zone of about 20–30 m. This inner zone and the buffer zone must be kept clear of debris such as dead leaves, fallen branches, rubbish, firewood etc. The outer zone includes the paddocks and wooded areas. Draw up a plan (from your property plan) and identify the different zones. This plan should include neighbours' properties also as in order to get to your property the fire will be travelling from them.

Any possible problems should be identified and counteracted. This may mean moving something or creating something. Fire risk problems include bushland, long grass, stacks of firewood, loose hay, certain trees and chemicals. Fire travels faster uphill so be aware of this. Pluses include dams, swimming pools, drives and laneways (which can be used for firebreaks, evacuation routes and for fire services vehicles), gravelled areas, short grass (perennial pasture is both good summer feed and fire resistant), ploughed land, deciduous orchards, vegetable gardens, arenas and windbreaks of fire retardant trees. Plan to place firebreaks in the areas that fire is most likely to come from and to reduce any fire risk problems in these areas (see p. 175).

A novel idea (if building horse facilities from scratch and if the lay of the land will allow it) is to put a ring laneway/exercise track around the inner zone as in a trotting/racing stable. This ring can serve many purposes including an exercise/riding ring, a laneway system to the paddocks around the property and a firebreak. Either side or just one side of the laneway can be planted with fire resistant trees. If it will be used

Figure 4.2 Fire hose and fire extinguisher Credit: Jane Myers

frequently it can be surfaced, if not it can be sown with hard-wearing grasses and can be strip/block grazed from time to time. Strip/block grazing with electric fencing will stop the horses from having their own race meetings.

Alternatively, if the local council requires that you have a firebreak around the outside edge of the property the same thing can be constructed there instead. Check what is required as some local councils, as well as insisting on having firebreaks, also insist on it being kept ploughed. However, you may be able to argue that this will lead to weeds and erosion and a surface or short grass will be better. In the case of a perimeter firebreak, the inner edge can be planted with fire resistant trees to further reduce fire hazard.

With both of these designs the driveway to the house will need to cross the 'track' which will need to be gated. These gates can be fitted so that they can be used both on either side of the driveway or on either side of the track.

Property maintenance for fire prevention

In the inner and buffer zones the grass must be kept short and, if possible, green. Any combustible material should be removed. The ground under all trees must be kept clear. Leaves should be cleared up regularly. Laneways and/or firebreaks should be grazed or mowed regularly to keep the grass short. Any remaining long grass should be mowed or whipper-snipped.

It is also a good idea to plan to get a petrol-driven pump for fire fighting because in a fire situation the power is often the first thing to go down. This pump can be situated next to the power pump with a connection to the same system (see p. 77). A sprinkler system around the inner zone will also be invaluable in a fire situation. Make sure all

Figure 4.3 Tree-lined driveway Credit: Jane Myers

connections are metal not plastic as these will melt in a fire. Equip buildings with fire extinguishers, and install heavy duty smoke detectors also. See Chapter 12 for resources for fire planning information.

Access

The property must be easy to get into, out of and move around in. Make sure that the exit from the property is located so that you can see in both directions of the road. Remember that if you are towing a float you will need more time and will pull away much more slowly than without a float. Likewise, turning into the property with a float will also require slow manoeuvring.

If the property is in a flood risk area, make sure the driveway is constructed to protect you from being stranded during flooding.

Plan to put the gateway at least the length of a car and float in from the fence line so that the vehicle does not have to be partly sitting in the road when you are opening and closing the gate. This will also be much safer if you plan to ride in and out of the property.

Figure 4.4 Laneways between paddocks Credit: Jane Myers

Be generous in your allowance for the width of the driveway. Either side of the driveway can be planted with bushes or trees for fodder and as a windbreak (see Chapter 9). Too narrow will be a problem for trucks (hay/water carters). The driveway can either go up one side of the property or somewhere near the middle. The latter involves more fencing but often looks nicer. You may even plan to have separate driveways for the stable and the house, although on a small property this will use a lot of space that could be used for pasture. This arrangement is more important if the property is to be used as a business, i.e. a riding school. Allow plenty of turn around at the end of the driveway for large vehicles.

You must be able to move around a property quickly and efficiently in all weathers, without having to go through one paddock to get to another. Plan to have laneways that link paddocks to the main facilities area. These laneways can serve several purposes. They can be planted with fodder trees on one or both sides. As well as providing fodder these trees will be windbreaks and will give shade. If the laneways are created along a contour they can help to divert water so that it does not run down the hillside (see p. 65). These laneways can also be grazed by horses from time to time (by quiet horses only, as long narrow areas are not safe if horses start to gallop). On smaller properties it is not usually necessary to be able to drive a vehicle on the laneways except in an emergency. They will need to be surfaced if they are used frequently by vehicles. The best location for laneways is on ridgelines where the ground will be dry, if not, avoid wet areas.

If you keep horses you should always have a closed gate preventing any loose horses from getting out on the roads. Depending on the layout of the property, it may be possible to place this gateway beyond the house and before the horse facilities and paddocks. This will eliminate the need to open and close a gate when entering and leaving the property by car.

Buildings and facilities

Now that you have identified the different areas of your block, you should be able to see one or more possible sites for the house if it does not already have one. When deciding on a site, consider such factors as stock and vehicle access, drainage, and wind protection. Avoid building in a floodplain. Sheds and stables should be located in elevated areas sufficiently large enough to be used as a holding area for horses in the event of a flood.

You will also have to consider the cost of providing the site with electricity, telephone lines and a water supply. Choose a site that has the best combination of all these factors. The ideal location for the buildings is in the centre. This allows easy access to all parts of the property; however, this must be balanced against the extra costs of a longer driveway and power/phone lines to the buildings. In practical terms this usually means that the buildings are located within the front half of the property, perhaps a quarter to a third of the distance in from the front entrance.

The house and other buildings need to be angled so that they receive maximum sun in winter and maximum shade and breeze in summer, protection from the wind and to take advantage of the view etc. You need to speak with neighbours (or the past owner) to find out which direction the bad weather comes from and the path that the sun takes at different times of the year. Buildings should not be positioned too close to the

Figure 4.5 Layout of a property Credit: Gary Blake

boundary as this creates problems if you need to improve drainage or to carry out maintenance at a later date.

The house can either be located in the same area as the horse facilities or separate to them. There are many advantages to having them close, including the reduction in costs for supplying power to both, added security due to proximity to the house and ease of use, that is saving travelling between the two when caring for horses. The disadvantages are possible smells and flies. However, if the property is managed well these can be minimised. The stables (if you are building them) can be situated downwind of the house but good stable and pest management should make this unnecessary.

A small but important point is that it is a good idea to design your facilities so that inexperienced/nervous family members/house sitters/friends can feed your horses without having to go into small spaces with them if you are not there. This will free you up considerably if ever you need to be away.

Stables and yards need to be positioned away from water courses such as creeks and rivers, at least 50–100 m away and at least 2 m above the highest water level for your area. Speak to your local council for advice. If the stables must be near water they must be sited on an impervious layer such as concrete or compacted limestone as manure and other debris should not be allowed to enter the waterway.

When building stables, there are many options available to you. They can be single, in rows, in a barn and so on (see p. 131). Every situation is different. Factors that will determine where you put them are drainage, nearest power, and distance from other buildings.

Plan to position stables so that they either have their backs to the wind (in the case of an open row) or so that the wind does not blow straight through from one end to the other (in the case of a barn). In the Southern Hemisphere horse stables and shelters should preferably face east and if this is not possible then facing north is the next best alternative.

Individual shelters in paddocks are not recommended unless they are enclosed by a yard in which case they become a sacrifice yard (a sacrifice yard is an area that horses can be confined in when the land needs to be rested for one reason or another) (see p. 115). Otherwise horses track to and from these areas causing compaction and other land degradation problems. Horses do not tend to use paddock shelters effectively unless they are forced to (i.e. when enclosed in a yard with one). On agistment properties separate paddocks and yards are usually necessary because people are very cautious about paddocking horses together. In this case position shelters (with yards) as near the gate as possible (this can be part of the entrance to the paddock) and confine horses there when necessary. The gate and the yard with shelter should be in the highest and driest area to reduce drainage problems. Surface the yards and they will become useful areas for handling, feeding and confining horses.

If possible, areas that are used for working horses should be situated so that they can be seen from the house and stables. This is so that people working horses can be checked on by other family members for safety reasons. This does not necessarily have to be in full view of the house, just somewhere that can be seen from certain parts of the house if necessary.

Remember to take your neighbours into consideration when planning, for example non-horsey neighbours might not appreciate stables being built near their house.

When planning the horse facilities allow space for a sacrifice yard of approximately 50 to 100 m^2 for each horse (see p. 115). Sacrifice areas are vital on a small property. They can either be attached to the stable complex (if building stables) or separate areas with either manmade or natural shelter/shade.

Planning water management

Planning water management includes managing the water that arrives on the property (rain water, creeks and rivers, bore water etc.) and its storage and distribution. It also includes managing the water that leaves the property, i.e. run-off and waste water.

At the planning stage you need to consider how you are going to collect and distribute water to the house and animals. Areas that are built on create much more water run-off than areas that are not built (water seeps into areas that are not built). This run-off water should be collected in tanks. As well as providing you with free water you will also be reducing the damage that water run-off can do if left to pour off the roofs (see p. 72).

You need to work out a water management plan *before* you fence, not the other way round. Water run-off from slopes should be directed across the landscape rather than down any hills. This can be helped by fencing paddocks and laneways along contour lines and keeping any fences that go down hills as short as possible (see p. 65 and Figure 4.1).

If the property already has a dam/s, plan to fence it off and if necessary reticulate the water around the property.

If the property is going to be set up for irrigation, a consideration is the distance from the power to the pump and the distance from the pump to the main irrigation network.

Avoid putting watering points near gates or in valleys as horses will traffic downhill to them creating the conditions for erosion.

Paddocks and fencing

On small properties a balance must be struck between having enough paddocks to rotate the animals and having too many small paddocks which are more susceptible to land degradation. Also smaller paddocks result in horses coming into contact with the fences more often. This increases the risk of fence injuries.

The shape of the paddocks will be dictated by the available space (after the buildings have been allocated space) and the landscape. Wet areas and dry areas must be separated so that they can be grazed at the appropriate time of year (see p. 91). Square paddocks are the most economical to fence but they are unlikely to fit in with designing your property to minimise land degradation if the property is undulating.

Bad subdivision can be the start of problems such as soil erosion. Whenever possible, aim to fence along the contour lines in order to reduce erosion (a contour is a line that connects points of equal height on the land). Ridgelines and valleys can be used as a starting point.

The fences themselves do not stop the water from travelling down the hillside, although they do prevent horses from tracking up and down. Having the fence along the contour lines means that trees can be planted on this line (on the other side of the fence in a laneway or between a double fence). This vegetation, including the longer grass that will grow between the trees, helps to slow water flow and to divert it along, rather than down, the slope. Where there is a greater slope it may be necessary to create a bank of earth on the top side of the laneway to divert water. Any cultivation to improve drainage and water uptake such as aeration can be done using the fence as a guide, thus keeping it along the contour line (see p. 102). Of course fences will have to go down some hills but aim to make these sections as short as possible.

Initially fencing the interior of the property with temporary electric fences is a very good idea until you have worked out the best place for more permanent fences (which can also of course be electric). On very small properties a mix of permanent (at least on the perimeter) and movable electric fences can work well (see p. 150). Portable electric fencing is very useful on horse properties as in small areas it is safer than solid fences and the layout can be changed very easily.

You will need several paddocks for your horses. Whenever possible, the horses should graze as a herd as this enables maximisation of the total grazing area by rotating paddocks which in turn allows paddock maintenance to be carried out. Another benefit of keeping horses in a herd is that they will settle and graze rather than spend time walking the fence line (causing land degradation and possible fence injuries) trying to get to other horses. Remember that horses are herd animals and need companionship to thrive (see p. 7 and p. 18).

Very small paddocks can be difficult to maintain with machinery. A ride-on mower is usually adequate for small paddocks and it can do light harrowing as well.

Paddocks with rounded corners are easier to maintain and safer for horses. A harrow or slasher/mower can get up to the edge of a rounded corner and fast-moving horses are guided around a rounded corner rather than into it. Never have acute angles in paddocks. These and any other corners in paddocks that are already fenced can be eliminated easily by fencing across the corner with electric tape or braid. This fenced-off area can then be planted with bushes or trees.

Plan paddocks so that there are no power line/telephone line guy wires inside the paddock. If this is not possible, either fence around that area or at the very least make the wires as visible as possible by attaching something white or bright to them.

Flora and fauna

You need to plan where the trees and bushes will be planted on your property. As well as in laneways and between paddocks (using double fencing), trees can be planted in the corners of paddocks (for fodder, shade and shelter) at the same time turning a corner into a curve making it safer for horses. Trees in the middle of a paddock provide shade at all times of the day. This, however, makes it harder to work around. Groups of trees will need to be fenced so that horses cannot ringbark them and compact the ground near to the trunks (see p. 182).

Any existing wildlife on the property needs to be protected and habitat created for more (the more varied the ecosystem, the more sustainable the property will be). This can be in the form of fencing off wetlands, leaving areas of natural vegetation (fenced off) and so on. Maintaining and creating habitat for wildlife has numerous benefits to the property as a whole (see p. 33 and p. 173).

Security of horses and equipment

Reduce the opportunity for theft by careful planning. Aim to have the tack room near the house if possible to reduce the risk of theft. A second driveway can decrease the security of the property as can gateways that lead from paddocks directly out on to roads. At the same time the property must be accessible to the emergency services in cases such as fire.

Figure 4.6 Private property

Credit: Jane Myers

5

Water supply and conservation

Water is the most important element, yet it is often overlooked when designing a horse property. How water is collected, stored and distributed is one side of the equation, yet the effect that horses and people have on the water system as a whole must be taken into account also. If, as landowners, you can make taking care of the water system something you *want* to do rather than *have* to do the experience becomes enjoyable, profitable and enhances your lifestyle. The sooner good environmental practices are put in place, the less damage will be caused to the water system and therefore the environment and it is easier to prevent that damage than to repair it. Good environmental practices result in reduced or eliminated water pollution, better horse health, lower costs in the long term, a better and more valuable property and a nicer place to live.

Managing and conserving water

Clean water is essential for healthy horses and healthy land. A horse needs access to clean drinking water 24 hours a day. A single horse drinks between 20–70 litres a day which amounts to up to 19 000 litres of water a year, depending on the location (heat and humidity), the work it does (more work equals more sweat), the type of feed (dryer feeds require more water for digestion therefore horses eating green grass need less and horses eating hay need more). When calculating water requirements for the property as a whole allow for evaporation, seepage and drought.

A poorly managed horse property will first have a detrimental effect on the local water system. This will lead to damage to the water system at large (water table, rivers and sea). The way in which you manage your land will directly affect your water quality and that of your neighbours. Even if your land does not have a creek, river or dam, what you do will affect water further on.

The natural system

Water comes from springs, rain or melting snow. It runs over the land or underground until it reaches the ocean. The area of land that catches rain and snow which then drains or seeps into the surface (creeks and rivers, lakes, dams or wetlands) or ground waters is called a watershed. We all live on a watershed and everyone has an effect on water.

As water runs towards the watercourses it passes through a riparian zone (in the natural state). The riparian area is the land that immediately borders and surrounds creeks, rivers, dams/lakes and wetlands. It usually supports a high level of biodiversity due to several factors such as deeper soil (from silt deposits), greener grass, water and shelter. Many species of plants and animals live in the riparian zone.

These areas are important because they filter sediments and nutrients that are washed off the land (too much sediment and too many nutrients disrupt aquatic life). They buffer the negative effects of floods and wind because their plants hold the soil together and decrease soil erosion while providing shade and shelter in the heat and cold to wildlife. They also provide a windbreak for your paddocks, yourself and animals if they contain large types of vegetation.

After passing through a riparian zone, water then carries on to a creek or river and eventually out to the sea or it is captured in natural or man-made lakes (dams) or wetlands, where it is held for a while before eventually moving on to the next stage (creeks and rivers).

A healthy riparian zone means that cleaner water enters the waterways – creek, river, lake, dam or wetland.

Wetlands in particular are vital resources because they act like a giant sponge soaking up water in times of high rainfall and releasing it slowly into the next stage of the waterways – creeks and rivers. This reduces the incidence of flooding and erosion. This is why areas that previously had large wetlands that have been drained for development often have flooding problems in times of high rainfall. Like riparian zones, wetlands act as a natural filtering system by removing pollutants as they pass through

Figure 5.1 A healthy riparian zone Credit: Jane Myers

the vegetation (the vegetation 'holds' particles and also uses them for growth when they settle). This means that when the water finally reaches a creek or river or the sea it is much cleaner than if it had not passed through a wetland first. Wetlands also provide vital habitat for wildlife (ducks and other waterfowl need places to care for their young and to shelter from predators; frogs need places to spawn; turtles need habitat; other animals rely on the inhabitants of wetlands for food i.e. birds of prey).

The negative effects of horse properties

Poorly managed horse properties affect the waterways in many ways including those listed below.

- Areas of bare soil and manure (these may be either relatively small areas or may even be totally bare paddocks) are blown away during dry periods, or washed away in wet periods. These are deposited in other areas including the waterways.
- Areas of compacted soil repel rather than absorb water. When it rains the water runs over the ground rather than soaks into the ground. This run-off takes soil and manure with it into the waterways.
- With some soil types, rain can also wash pollutants into the soil and pollute ground water.
- Rain run-off from buildings that is allowed to run near manure heaps and over horse yards, carries manure with it into the waterways.
- Fertiliser added to the land can also end up in the water system.
- Chemicals (shampoo etc.) from a horse wash can pollute the waterways.

All of this means that nutrients, bacteria, viruses and parasites enter the waterway and contaminate and pollute the water, causing big problems in the water system as they cause algae and aquatic weeds to form. As these grow they shade the native plants in the water causing death to these plants and aquatic life. Eventually all of these contaminants end up in the sea where they pollute marine life (flora and fauna) of all kinds.

Figure 5.2 Drinking from a dam Credit: Jane Myers

In addition, if horses are allowed to have direct access to water sources and riparian areas they have a damaging effect on them.

- Mud is created which will pass on into the water system. Soil (silt) in the water can clog fish gills, cover spawning beds, smother fish eggs and make it hard for fish to see their prey.
- Horses dung and urinate into the water causing pollution, even very small amounts of urine (ammonia) can be toxic to fish. Nitrogen and phosphorus are the biggest culprits for stimulating algae blooms.
- Horses will eat and trample seedlings and vegetation, which reduces the habitat for wildlife that relies on them. Vegetation around waterways prevents nutrients from entering the system and causing problems therefore no vegetation leads to big problems.
- Horses compact the wet soils around the edge of water zones which suffocates plant roots and causes serious damage.
- Bare soil results as horses pull out plants including the roots, which leads to erosion (plants' roots hold soil together).
- Erosion changes the course of creeks and rivers which leads to further erosion as the water moves faster (rather than meandering) taking more soil with it.
- Dying plants and algae give off unpleasant odours and cause a drop in oxygen levels in the water which becomes a downward spiral.
- Contaminated water causes problems if it is used as drinking water for animals or humans. Swimming in contaminated water can be a health problem.

For all these reasons, allowing large grazing animals to have access to water zones compounds the problem of poor management.

Signs of damage

Some of the signs that domestic stock has already caused damage to riparian zones and waterways are:

- dying trees and saplings
- short vegetation
- scum and algae on dams and still water
- smelly water
- dying fish
- lack of wildlife
- plagues of mosquitoes
- pugging on the edges of waterways
- bare areas on the sides of waterways.

Managing waterways

Clean water starts with either protecting or creating riparian zones. This can be done by fencing waterways and the riparian zone, preferably permanently, from domestic stock.

Fencing off these areas enables the buffer zone to grow undisturbed. Initially this can be a simple electric fence. Later the area can be fenced with more permanent materials as well as electric fencing on the horses' side, as the vegetation that grows on the water side of the fence will tend to keep shorting it. Domestic animals will also try to reach the grass on the other side because, as we know, the grass is always greener on the other side of the fence. Take care to construct your fence in such a way that native animals can still access the area.

Speak with your local Landcare group about the recommended distance that the fence should be from the waterway – 8 to 16 m is the minimum that you should be aiming for. You may be able to get either financial or physical help from your local Landcare group.

This fence will allow grasses and trees to grow naturally. It may be necessary to re-vegetate these areas if they are damaged to the point that they cannot re-vegetate themselves or weeds have taken over. If weeds are a big problem care must be taken that their sudden removal does not result in further degradation. You may have to graduate weed removal with introducing more desirable species. Find out and obtain the right plants for your area but remember you must fence first.

Once restored, this wetland will be a far more beautiful feature on your property and will operate far more efficiently than simply providing grazing for stock.

Build water crossings and water access points if necessary. A crossing should be fenced on either side and concreted or covered in rocks. Access can be created by fencing a 'u' or 'v' shape into the creek or river to give some access and the area covered in stones to reduce pugging and silting, but it is much better to prevent access altogether. If necessary the water can be pumped (check with the local authority if it is a creek or river) to troughs or holding tanks and distributed from there.

Locate paddocks and yards (anywhere where manure will run off) away from waterways and wetlands. Locate or store manure heaps in areas where water run-off does not run through or pass nearby. This can mean that you have to install rain gutters and either collect water in tanks or divert the run-off. Do not use steep hillsides to paddock horses as the manure is washed more readily into the water system. Horses on steep land also cause erosion that will lead to more soil ending up in the waterway. Don't allow bare areas to form anywhere on the property, bare soil will be blown or washed away and is usually deposited in the waterway.

Site any new buildings and yards as far from a waterway as possible (aim for at least 100 m, again you will need to check with your local council) and on the highest parts of the property. Make sure that contaminants do not leach into the soil by situating them on an impervious layer such as concrete or compressed limestone.

Water sources

Water primarily arrives on your property via rain that lands directly on the land as well as rainwater run-off that travels from neighbouring properties, across your land, towards a creek or river. Capturing this water and storing it in tanks and dams for later usage is the most efficient way of ensuring self-sufficiency in terms of water.

Other potential water sources include 'town' water (water that is piped to your property by the local council in urban and some semi-urban areas), bore water which is water that is pumped up from the underground water table, natural spring water and

Figure 5.3 Water tank Credit: Jane Myers

creeks or rivers that run through or alongside the property.

All properties have the opportunity to collect rainwater both directly and as run-off.

Tank water

Rainwater from the roofs of buildings should be collected and stored by fitting gutters and spouts to all buildings which run to storage tanks. Any water that is surplus should be channelled straight to a dam or creek without having to go through a polluted area first (through horse yards and past manure heaps for example). If this tends to happen on a regular basis there may be a case for putting in more tanks.

From the storage tanks the water can then be pumped to where it is needed. If the property is hilly an extra tank can be positioned on a nearby hill and water from the tanks near the buildings pumped to this (header) tank. The header tank then gravity-feeds troughs and wherever else the water is needed, the higher up the hill the more pressure it will have. Having a header tank creates more space in the tanks near the buildings so that the next time it rains they can be filled again. This system means that instead of a pump operating every time a tap is turned on, the pump only has to operate infrequently to pump the water uphill to the header tank. Another system involves pumping water from a dam or creek or river to a header tank and then this water is used for toilets, washing horses and so on, saving the clean tank water for drinking and washing (people).

If the water in the tanks runs out and there is no other way of replenishing it then water will need to be bought. This will be delivered by tanker and it works out relatively cheaper to have a larger tanker deliver the water as most of the cost is in the delivery.

Dams

Dams are another valuable way or catching and storing water that would otherwise run off your land. As well as being a useful supply of stock water, dams can be an attractive recreation area and habitat for wildlife. This wildlife will also be beneficial to the property by reducing pests. A dam can be stocked with native fish, in particular minnows which eat the larvae of mosquitoes. Frogs eat lots of pests so encourage them to live in your dam by creating areas for them to live in (under rocks).

Dams must be installed by a competent earthmover to reduce the chance of failing. They also have to be managed properly if they are not to become stagnant toxic pools.

On a small property it is better to have one large dam than several small ones if space is limited, as small dams do not work as well. A dam needs to be large and deep (at least 4 m) in order to keep the temperature low and reduce evaporation. Small dams stagnate quicker due to increased evaporation rates.

Figure 5.4 A fenced dam before revegetation Credit: Jane Myers

Check with your local council before having a dam constructed, as you may need permission. Expert advice should be sought also as some soils will not hold water (such as sandy soil with no clay). Clay can be brought in but it adds to the expense. The dam must be compacted properly when constructed, using heavy machinery, and the spillway properly placed and effective. If a dam fails and causes damage to a neighbour's property you may be liable for the damage.

It is best to fence off a dam from stock and pump the water when needed to a tank and then reticulate to troughs. A buffer zone should be created around the entrance to the dam for all the same reasons that are outlined in the earlier section on managing waterways.

Water should be taken from the middle levels of the dam. If water is drawn from the top 200 mm of a dam, it is more likely to be contaminated with potentially harmful micro-organisms. However, water taken from near the bottom of a dam, will be colder and lower in oxygen due to the action of micro-organisms that use oxygen to break down any organic matter on the dam floor.

Figure 5.5 Horses have access to this dam behind the fence; notice the bare banks. In the foreground, access is restricted and the banks contain healthy vegetation

Credit: Jane Myers

One method of cleaning a dam that has an algae build-up is to put a few bales of hay (still in their strings) into the dam which float and the algae sticks to them. They can be removed after a few weeks. Lucerne hay seems to work the best. As well as helping to clear the algae, they provide a floating platform for wild birds. Periodically, just before or during a wet period it is also a good idea to siphon or pump water from the bottom of the dam out on to the paddock. This results in fresh oxygenated water replacing low oxygenated water from the bottom of the dam.

A dam should preferably be shaded which will also reduce evaporation. This can be from nearby trees and from water lilies. However, trees should not be planted into the banks as this will cause the dam to fail if the trees die, as the roots will shrink and cause leakage.

Town (mains) water

Properties in suburbia have town (mains) water. However, this would usually be used for domestic purposes only. Watering stock with mains water is an expensive option, plus it may have additives such as chlorine that will be unpalatable to the horses. If your property has mains water, plan to also install rainwater tanks on any buildings to catch rainwater run-off that can then be used for the animals. Mains water is usually either full pressure or trickle feed (in some areas). The latter requires a holding tank as it is a constant trickle to your property (rather than full pressure) and is stored in a tank which you then pump to where it is needed.

Bore water

Bore water is water that is drawn up from the water table. This is the natural aquifer that is underground. In some areas the water table is huge and contains abundant good quality water. In other areas it is either non-existent, limited or poor quality. Aquifers also vary in how close they are to the surface. In the past people have tended to use bore water as if it is a never-ending water supply, however, it is being used at higher rate than nature can replace it, so use it carefully.

Drilling companies will often do an initial inspection for nothing. However, the next stage may cost several hundred dollars as the contractor uses records and local knowledge to tell you the chances of finding water. Having a bore completed (fitted with a pump etc.) will usually cost several thousand dollars. In many areas there are now restrictions on putting down both domestic and irrigation bores. Check with the authorities before installing a bore.

Bore water can be used for stock. Remember that it should be tested first. In some cases it is also suitable for human consumption. If you are thinking of installing a bore on your property you need to weigh up the costs with the benefits. Some areas have very little ground water (or poor quality ground water) that may be very expensive to reach and may be poor quality when you reach it. It may even run out at some point. Before going ahead, speaking with neighbours about their bore experiences can save time and expense. Some areas, due to reasonably high rainfall, do not need bores as adequate amounts of water can be collected as rainwater run-off (and stored in tanks and dams). With a bore you may still need a tank for storage and distribution and may need filtration equipment.

Natural spring water

Some properties have spring water which is where underground water naturally comes up to the surface. This is obviously highly desirable as it should mean that the property will always have water. Some spring water can be quite heavily mineralised which may make it unpalatable, at least initially, until the animals become accustomed to the taste. A dam can be created around a spring which results in a constantly full dam.

Creeks and rivers

Domestic animals should not have access to creeks and rivers for drinking water due to the extensive damage that they can cause to the water system. In addition, if a creek or a dam has a sandy bottom the horses will take this up each time they drink which can lead to sand colic (due to an overload of sand in the gut). Another problem is that if creeks and rivers are not running they will become stagnant. If the property has a licence to harvest water from the waterway it is better to pump the water to a dam, troughs or holding tanks.

Using water

Irrigation

Irrigation may be installed for several reasons, including:

- the growing season for pasture can be extended
- grasses that could not normally grow can be grown
- pastures look better
- water is available for fire control.

However, irrigation is not essential on a small property and it is quite possible to manage without it. Installing irrigation is expensive in both monetary and environmental terms.

If your property already has a system in place then your main concerns are that the system is operated efficiently without water wastage. There may be things that you can do to improve the system as shown below.

- Make sure the system is not putting more water on the ground than it can cope with. A maximum of 10 mm per application is recommended to avoid fungal problems and waterlogging. However, if the soil is compacted any amount of water, no matter how much or how little, will run off without soaking in.
- Make sure the grasses and legumes are of a type that will benefit from irrigation, i.e. perennial rather than annual. Either sow grasses and legumes that will benefit from the water or do not water. There is no point in watering plants that will not grow at that time of year or watering a paddock full of weeds. Incorrect irrigation can actually bring on more weeds.
- Keep in mind that some grasses cannot cope with the sometimes higher levels of salt in bore water.
- Carry out paddock management (i.e. paddock rotation, manure management) as for non-irrigated pasture. Rotations will need to be more frequent due to the herbage growing quicker.

- Aim to be careful with the water whether it is from a dam or a bore. Minimise water usage by watering at the right times, do not irrigate during the day in hot weather, early morning or later in the evening is best. Use timers so that your system does not get left on and waste water.
- Before the dry season check and maintain the system. Protecting sprinklers from horses is difficult but must be done.
- Do not allow water from irrigators to pump onto roadways or other unpastured areas. This is wasteful and upsets neighbours who may not have access to bore water.

Water for irrigation comes from dams or ground water via a bore (mains is not an option due to the expense). On small properties there would not usually be a large enough dam for irrigation. If you are using bore water it may seem as if you have access to endless water but this may not be the case. Bores can dry up, especially if too many people in the area are tapping into the same source.

If the property does not have a system in place, you need to weigh up the cost of installation and possibly sowing different pasture. Perennial pastures can be watered with good results, but keep in mind that pastures containing a high percentage of annuals will not benefit from irrigation. Watering this type of pasture after the growing period has finished and the plants have set seed and gone dormant until next year will not cause them to start growing again. Also, if you water these types of pasture you may set up conditions for weeds to get ahead of grass.

When all of these factors are taken into account it may be that it is not worth the expense. Weigh up the costs involved (calculating the chances of failure of the system, i.e. the bore may stop producing water) compared to the costs of using sacrifice yards (see p. 115) and buying extra feed. Remember that pasture can often be vastly improved even without irrigation for a fraction of the cost. Even when the shortfall in feed has to be made up by purchasing feed it can still be far lower than the cost of irrigation (including the environmental cost and the labour cost).

If a new irrigation system is installed, aim for ease of use and economy. In small paddocks it can be positioned around the fence line. A portable system using hoses and freestanding sprinklers can be used for the middle sections if necessary. In larger paddocks it will be necessary to situate permanent sprinklers in the middle of the paddock as well as around the edges. These can either be in the form of pop-up sprinklers – which can work well with horses – or sprinklers fastened to posts to protect them from horses. If placed high above the ground (i.e. 1.8 m (6 feet) above the ground) they seem to attract less damage from horses than if they are set lower. Where possible these should not cause obstructions, for example to machinery used for paddock maintenance.

If you have no experience with irrigation systems, talk to people who have already installed one to get some ideas. Some irrigation suppliers give a design service – however, this will be aimed at using all of their own products whereas it may be more cost effective to use products from various outlets. Also make sure that the design is not more elaborate than you need.

Reticulated water systems

Water can be reticulated around the property from holding tanks. This can be done using polypipe that runs to troughs, automatic waterers in stables and yards and also to sprinkler systems if required. Unless you are using a header tank (see p. 72) you will need a pump permanently situated near the tank. An electric pump will automatically start as and when required without you having to do anything.

If you are relying on an electric pump, it is also a good idea to have a petrol pump that can quickly be attached to the same system in the case of fire. When a fire occurs, the power supply is often the first thing to go down rendering you unable to fight fires. A portable petrol pump is useful in that it can be moved around the property if necessary.

A reticulated water supply usually relies on a float-valve to control the water-flow into the trough. These troughs can be made out of concrete, plastic or steel. If the float valve is a type that horses can tamper with, it should be covered to prevent this happening.

Trough sizes can range from small plastic or metal 'automatic drinkers' (usually in a stable or yard), to larger troughs for the paddock. A round trough with up to 1 m diameter or a bath-sized rectangular trough is ample for most small properties. Troughs can be made by a handy person out of a recycled plastic drum (feed stores sometimes sell these) cut in half either down the length or across the middle. With the former a float can be fitted and covered making an automatic trough. Old baths are commonly used as troughs on horse properties particularly if the water supply is not automatic (i.e. just using a hose attached to a tap) or again a float can be fitted. Baths are easy to clean and are fine as long as they have no sharp edges that will cause injuries.

Avoid placing troughs in corners of paddocks where some horses may be in danger of being cornered by other horses and unable to escape. Troughs can be placed so that they can be accessed from two paddocks to save costs (unless you have a double fence). Troughs should be set on a level sand base, surrounded by a firm surface that will resist hoof action. Place troughs in an area where the ground drains well in wet weather with gravel set around to reduce mud and dust. Because this area will be bare of grass, also consider positioning troughs on the windward side of the stables or other buildings where they will protect buildings and be most useful in a fire situation. Plant fire resis-

Figure 5.6 A concrete trough

Credit: Jane Myers

Figure 5.7 Bath with float Credit: Jane Myers

tant trees in this area also and the animals will tend to stand there on hot days when there is an increased danger of fire.

The diameter of the water pipe leading into the trough should be of sufficient size to allow refilling of the trough in an acceptable time. Typically 32 mm is a starting point in a non-mains pressure type reticulation system.

Whatever the water supply, it must be checked daily (more often in hot weather) as it can be knocked over, become contaminated (i.e. by a dead rodent/bird or manure), the horse may not be able to get to the water for some reason (i.e. someone accidentally shuts a gate), troughs can break down, creeks can stop flowing and water can freeze in cold areas.

Conserving water

As well as protecting waterways, there are many things that you can do to conserve water. Small amounts of water add up meaning that not as much water needs to be taken from the water system as a whole. Some of the things that you can do to conserve water are suggested below.

Sponge horses rather than hose them down as hosing uses lots of water. Aim to wash horses less often and groom more often. Your horse will prefer that anyway.

If you do hose, either hose your horse on a grassed area or have the water drain into a system for recycling, such as a grey water system that is already attached to the house. If enough water to warrant it is being used, the stables can have their own grey water system.

Other water-saving suggestions include:

- either set up your property so that it does not rely on irrigation or use irrigation carefully if it is already installed
- plant grasses and crops that use less water
- reduce lawn areas if you normally water them and mow them
- water plants by trickle systems or micro-irrigation and only water in the early morning or late evening
- mulch plants to reduce water evaporation (this also keeps them warmer in winter)
- ensure taps and auto waterers do not drip (including paddock waterers)
- collect rainwater as much as possible
- be careful not to allow troughs to overflow when filling them.

When washing horses, look carefully at the products you use as some shampoos actually stop water from entering the soil. Use environmentally friendly products that are mild and biodegradable.

Speak to your local Landcare officer or Department of Agriculture contact who can help you with water saving and collection strategies.

Water problems

A common problem is that there is often either too much water or not enough of it. In many regions water arrives all at once and then not at all for very long periods (especially in Australia). By managing water correctly these slumps and booms can be ironed out to some extent.

Waterlogged land

Land becomes waterlogged when the soil is unable to absorb the water it receives. This can cause land degradation and health problems for your horses (i.e. mud fever, greasy heel etc.). It is very important that waterlogged land is managed properly otherwise the problem will become worse (see p. 91).

Flood management

Floods are more common than people tend to think and therefore they often have a false sense of security. If your property is situated in a 100-year flood plain this means that there is a 1 per cent chance of flooding per year or a 30 per cent chance in the lifetime of the average mortgage.

Floods can occur from floodwaters moving down a river or creek, which causes slow rising floods. Heavy rain or melting snow can cause flash floods. Floods can also result from a dam or levee bank failure.

Your local council will have information on floodplains in your area. Utilising Landcare principles and complying with building regulations will reduce some of the high cost of losses due to flooding. Forward planning will greatly reduce the amount of animal and human suffering in any emergency situation including floods. See Chapter 12 for resources on flood management.

Drought management

When a drought becomes prolonged, your horse must be kept off the land. This is due to the damage that horses can cause to dry soil. During a drought, hot dry winds will carry away soil and after the drought has broken the rain can cause a lot of damage if the paddocks are bare. It is wise to aim to lock up the paddocks while they still have 5–8 cm of growth on them for protection, otherwise land degradation will result.

It is thought that a drought can induce toxic conditions in certain plants, this is usually due to a rust or fungus that occurs when the plants have been stressed. Contact your local agriculture department to find out which ones, if any, are likely to be affected in your area.

Feeding horses during a drought obviously becomes very expensive. In a prolonged drought hay may start to become relatively more expensive than hard feed (grain and other concentrates). This can result in horses being fed a diet that is too high in energy for the amount of work they are doing, so always try to keep as much roughage in the diet as you can afford.

Bare soil equals dust in a drought so you are in danger of losing your precious soil to the wind and later the rain. Cover any bare patches in paddocks and along the fence lines with mulch of one kind or another. Possibilities include old hay/straw and composted manure.

In a drought you must check the water supply more frequently as it is more likely to become stagnant.

Once the drought breaks you will still need to take extra care of your property and horses. Even though it is very tempting, do not allow the horses out as soon as the paddocks show a bit of green as they take time to recover. This initial bit of green in the paddocks will have very little feed value anyway. Even when the grass starts to grow well, start off with limited grazing and build it up as the horses will gorge themselves at first. See Chapter 12 for resources on managing horses in a drought.

Impure water

As well as bacteriological testing, water can be tested for all manner of impurities.

Although horses can drink 'hard' (usually due to large amounts of calcium) water – once accustomed to it – magnesium salts are the first to reduce palatability (and therefore the amount of water consumed), so it is recommended that 200 mg/litre of magnesium be the maximum allowable amount. Two other factors that affect water quality and can be tested for are pH (acidity and alkalinity) and electrical conductivity (salinity).

A level of not more than 50 *E. coli* colonies per 100 ml of water is recommended. This is just a guide to the safe upper limit of bacterial infection of water for livestock.

Figure 5.8 Impure water Credit: Jane Myers

6

Pasture management

Australian and New Zealand horse owners are in an enviable position because horses can usually be grazed on pasture all year round. But many landowners, Australians in particular, do not make full use of pasture; paddocks are often regarded as somewhere for horses to exercise rather than as a valuable resource. In New Zealand the attitude tends to be different and pasture is utilised to a much larger extent for horses. In the past this underutilisation of pasture in Australia has been due in part to a belief that grazing in paddocks can cause a range of nutritional disorders in horses.

Managing pasture

Pastures often get neglected due to overstocking and lack of land care. Many horse paddocks have bare patches, weeds, rank grass, erosion and compaction. If not properly managed, the paddock can become muddy, dusty, sandy or eroded and can cause respiratory problems, colic, thrush, laminitis/founder, obesity and loss of condition.

As a landowner it is important that you maximise the potential of your grazing land (not including other areas such as remnant vegetation) for two reasons. The first is to maximise production of a cheap nutritious pasture yield and the second is for the land-care benefits that result from healthy vigorous pasture.

Good pasture is very beneficial (see 'The benefits of pasture', below). Badly managed pasture is an eyesore to all and creates degradation to the environment, through contaminated run-off, erosion, dust, loss of habitat for wildlife, and water pollution.

It is not difficult, and it is very rewarding, to improve or maintain your pastures and it adds considerable value to your property. Horse owners need to become more aware of the condition of their pastures and be alert to problem areas before they become expensive to fix.

Once a pasture has been improved it must be managed if it is not to deteriorate again. Once you have the knowledge you can do so much to improve and maintain your pastures using your own labour, hired labour and your horses.

The benefits of pasture

An Australian survey revealed that the average cost of feed for a horse each year was $880 (1998) and was the largest amount spent on the horse. By utilising any available pasture this feed bill can be reduced or even eliminated altogether. Well-managed pasture is an excellent feed source and provides a cheap, convenient and balanced ration for most horses. Pasture has been calculated as costing one-tenth of the cost of the next cheapest feed, pasture hay. Supplementary feed need only be given to certain horses when pasture is scarce, or the horse is in heavy work, is experiencing rapid growth, or lactating.

Other benefits of pasture include better health of your horses – the lungs are healthier due to the fresh air (as opposed to the ammonia and other airborne pollutants that a horse is forced to inhale when stabled) and the lowered head grazing position maintains the drainage system of the airways naturally.

When a horse is allowed to graze it is following a natural pattern of eating. These horses do not develop the behavioural disorders seen in confined horses. Even horses that have already developed behavioural disorders (from previous confinement) show a reduction in these behaviours over time when living a more natural lifestyle. The importance of browsing and foraging for horses cannot be underestimated as time spent at this activity forms the largest part of the day in a wild horse.

Horses thrive on a high roughage diet. Without it the gut cannot function properly (much more so in a horse than an omnivorous human) and the horse is plagued with gastrointestinal problems such as colic.

Horses kept in herds of two or more horses at pasture can interact with other horses naturally and benefit from their companionship. Pastures are far easier to manage if horses are rotated in groups around the various paddocks than if each horse has its own paddock.

Horses at pasture are exposed to sunlight so that they are able to synthesise vitamin D. They are also able to exercise freely.

The advantages for the owner are many also: keeping a horse at pasture saves on bedding and stable chores; the owner has to adhere to a less strict timetable; and the horse does not necessarily have to be exercised every day. In this way, time spent with the horse can be 'quality time', i.e. riding and training rather than finding that there is no time to ride because stable chores take too long.

For all these reasons, the pastured horse generally has a better quality of life than its stabled counterpart. For many owners however a compromise has to be reached between confining horses due to not having enough pasture and allowing horses to graze, yet this is still far better than no grazing or turnout at all. The priority should be to maximise the amount of pasture available (which may mean having to initially reduce your horses' turnout time) so that in the future your land is producing pasture to its full potential. It will then be possible to allow the horses more turnout on paddocks that contain good pasture which will result in a win/win situation.

Meeting nutritional needs with pasture

A mature horse requires approximately 1.5 per cent of its body weight per day in dry matter (DM: food with the water content taken out). Pasture varies in its DM content depending on such factors as the stage of growth (new grass is lower in dry matter as it contains more water). As a general rule a 500 kg horse needs 7.5 kg (DM) per day. It will eat more if it is available (it may eat as much as 2–3 per cent of its body weight if given the opportunity).

Good pasture will meet the nutritional needs of mature horses at rest and in light work. At certain times of the year (spring in some areas, the wet season in others), when pasture is growing quickly and has a higher nutritional content, pasture will feed pregnant mares and young growing horses over one year old. Lactating mares and weanlings may need supplementary feeding at certain times of the year such as in the winter or dry season.

Pasture yield varies between 80 kg of DM per hectare per day in a high growth period down to 10 kg or even less of DM per hectare per day in winter/dry periods.

Pasture planning

Some of the questions that you need to ask yourself when planning your pastures are:

- How many horses do I have and how many do I plan to have in the short and long term? (be honest)
- How much time and money do I have available to me for pasture improvement? Improved pasture will carry more horses than unimproved pasture.
- Am I planning to set up irrigation if it is not already irrigated?
- What effect will the local climate have on the pastures, i.e. how often does it rain/how hot and dry does it get?
- How many acres of pasture do I really have access to when I take off the area that the buildings, horse facilities, fenced-off waterways, laneways etc take up?
- What is growing in the paddocks at the moment? Do I want to change that?
- How much do I want to use the paddocks? Will it mainly be for recreation or nutrition or both?

Planning your pasture means that you can make informed decisions. Planning will involve taking into account that in order to have long-term productivity you may have to restrict the horses' access now and spend time and money on the pasture (and on supplementary feed for the horses). Horse owners who are also land owners need to learn to evaluate and monitor pasture in much the same way that a farmer does.

Stocking rates

The stocking rate is the number of animals a particular piece of land can support without land degradation. Because horses can be kept in a range of management systems, which varies in intensity from 100 per cent paddock-kept to 100 per cent confined with many variations in between, the stocking rate is just a guide which becomes more important the more you plan to rely on pasture as a feed source. If it is planned that horses are to be kept in an intensive management system (stabled/yarded all or most of the time), a much larger number can be kept on a given area. However, this is not desirable for the wellbeing of the horses.

It may also depend on local regulations; you need to check with your local council. Some councils have regulations about how many horses can be kept on a given area, some regulate over smell, manure management, flies and drainage. Increasingly more and more local councils will adopt policies for control. By managing your property well and involving other like-minded people who live in your area so that they also do the same, you may be able to ward off the time when the local council dictates what you do.

On average a horse needs approximately two to three acres of productive land. The feed potential of a property can be calculated once the actual grazing area is known along with knowledge about the soil fertility and type, type of pasture and the annual rainfall. When calculating your stocking rate remember that between one and two acres will be taken up with your buildings including the driveway, dams, stables, yards, arena and so on. Remnant vegetation should also make up part of the property. What is left over is what you will have available for grazing. Bear in mind that some areas will only be able to be grazed at certain times of the year, i.e. some areas may be wet/waterlogged for part of each year and fragile areas (such as sandy soil) will need more protection when hot and dry. Once the feed potential is known you can determine how many horses the area will support. Also, if an area is calculated as being able to support three horses full-time, it can support six horses for half of the time and so on.

The stocking rate is calculated as the Dry Sheep Equivalent (DSE). This is a figure that is used as a control for all agricultural land whether it is for sheep, cattle or horses. A dry sheep is an ewe (that is not pregnant) or a wether (a castrated male). An average 450 kg horse has been calculated to be equivalent to ten dry sheep. This of course varies with smaller or larger horses, the amount of work that they are doing and their breeding status (i.e. pregnant or not pregnant). The stocking rate of an area can be obtained from your local Department of Agriculture office. It is calculated taking into account the soil type and rainfall for that area.

Once the DSE is known, you can work out what your property should be capable of holding on a yearly basis, assuming that your property is managed well. For example the DSE for your area might be 12 DSE per Hectare (Ha) and you have four hectares of available land. This means that your land should be able to support 48 dry sheep or about four to five horses on a full-time grazing basis.

The actual number of horses that any area of land will carry varies enormously depending on many factors such as the soil, climate, pasture species, management strategies etc. Unless your pasture is already producing grass at a high level then the stocking rate can be improved with good pasture management.

Stages of growth of pasture

After the opening rains, grass starts to grow again but you must hold your horses until it becomes established. Putting them on it too early will give them little benefit and will not allow the pasture to grow. Plants get their energy from sunlight through their leaves so if they are grazed before the leaves have had chance to grow they cannot grow further. It will be roughly four to five weeks, or when the majority of the grasses have three full leaves, before the pasture can be grazed. Meanwhile you will need to continue

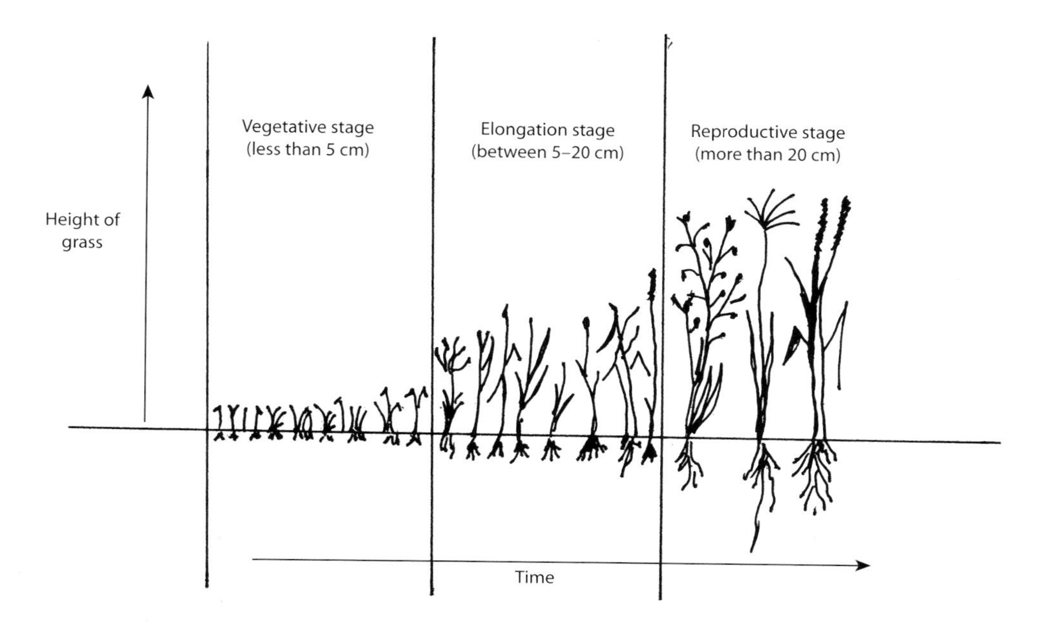

Figure 6.1 Stages of growth of pasture

Credit: Jane Myers

to use your sacrifice areas or use a paddock that you plan to renovate later that season. Allow the horses in this paddock for a few hours a day until the other paddocks have fully regenerated.

Plants have three stages of growth that you need to be familiar with in order to manage pasture efficiently. Stage one is called the 'vegetative' stage and it is when the grass is up to approximately 5 cm high. In this stage the plants do not have enough leaf area to trap the available sunlight and convert it to energy for growth. If grass is not rested at this stage it can kill the plant as it cannot cope with grazing pressure. If the plant is rested at this stage it grows new leaves and enters the next stage of growth.

Stage two is called 'elongation' as it is the period when the plant has enough leaf area to store and use energy. The stem grows taller but the number of leaves stays the same. This stage is when the plants are at their most nutritious and most able to survive grazing pressure.

Stage three is the reproductive stage when the seed head develops and pollination occurs. At this stage all new growth ceases and the plant becomes less nutritious to grazing animals.

Part of good pasture management is manipulating grazing systems to ensure that grasses stay in stage two for as long as possible. This is done by removing horses from pasture when the plant height is reduced to an average of 5–8 cm and allowing grazing again when the plants reach an average height of 15–20 cm.

There will be times of the year when it is difficult to keep all of the pasture in stage two of growth at the same time and then you will need to either mow it, conserve it (hay/silage) or let it grow, set seed and then either mow it or use it for grazing (albeit low energy) later in the season.

The length of time that it takes a pasture to recover during a rest period varies with climate, time of year and grass species.

Seasonal variations in pasture quality

Pasture quality and quantity varies markedly throughout the year. This will influence how much supplementary feed is required for pastured horses.

Different areas of Australia receive their rainfall at different times of the year with the temperate regions having wet winters and subtropical or tropical regions having wet summers. In New Zealand rain falls fairly evenly throughout the year and does not show the marked variation that occurs in Australia.

Pastures need a certain amount of warmth and water in order to grow. In some areas (particularly in Australia) there can be prolonged periods of little or no growth even in a 'normal' year.

These seasonal variations can be manipulated to some extent by using different grass and legume species. Generally speaking, grasses thrive in the cooler months and legumes do well in the hotter months. However, there are exceptions to these general rules.

Soil

Good pasture starts with healthy soil. Well-balanced soil will reduce the need for adding lots of supplements to your horse's diet as correctly balanced grass will do the job for you. You can seed land and add water but unless the soil is correctly balanced grass will not grow healthily. Healthy soil is necessary for healthy plants and therefore healthy horses. Healthy plants help create healthy soil, and vice versa. Most grass plants get their nutrients from the uppermost layers of the soil. Legumes reach further down into the subsoil.

Soils have profiles, this means it has layers. These layers have different names and different functions. The soil we generally see is the topsoil. This is the layer on the very top and it varies in thickness from a few centimetres to about a metre. This layer of topsoil may be covered in organic litter such as leaves. The layers underneath are subsoils, which are on top of weathering parent rock and finally there is bedrock. These layers vary in thickness and type giving a variety of soil conditions in different areas. Within a property boundary there is often more than one type of soil.

The main nutrients that the soil provides to plants are nitrogen (N), phosphorus (P), potassium (K), and sulphur (S). Collectively these are often referred to as NPKS.

The other elements that soil needs to provide are calcium (Ca) and magnesium (Mg). Plants also require trace elements such as copper (Cu), manganese (Mn), zinc (Zn), iron (Fe), aluminium (Al), boron (B), molybdenum (Mo), cobalt (Co) and selenium (Se).

All of these elements need to be present in the correct amounts otherwise an imbalance is caused. This will result in poor plant health or death of the plants. Also, if certain nutrients are missing then this may result in other nutrients being unable to operate. Nutrients become depleted from the soil in many ways including leaching, making hay that is sold off the property, grazing horses, collecting manure rather than spreading it and soil erosion.

In addition the soil requires good bacteria (which occurs when the soil conditions are correct), organic material such as decomposing plants/roots, animal matter such as dung and dead animals, micro-organisms, materials excreted by roots and by hyphae of fungi (the last two in particular 'stick' soil particles together to form aggregates), water, air and organisms such as earthworms and dung beetles.

Soil texture and types

In order to improve soil, its structure must be enhanced. This involves improved aeration, water infiltration, organic matter utilisation and nutrient utilisation. (Good soils tend to have darker topsoils as the utilisation of organic matter darkens the soil.) Soil texture is determined by the size of the mineral particles it contains. Particles range from gravel, to sand to silt to clay. The amounts in which these particles occur give the soil its texture. This texture is very important in the management of land.

Sandy and clay soils are at opposite ends of the spectrum and understanding how they behave goes a long way towards understanding their management. Basically sandy soil will need protection from horses in dry weather and clay soil will need protection in wet weather. Good soil is a combination of sand and clay because these contain the best of both worlds. These soils are called sandy/loam and loam soils.

Soil structure is a delicate balance that can be hindered by excessive pressure from horses (through compaction and pugging for example) and from excessive ploughing and cultivation which break up the soil aggregates and disrupt microbes in the soil until the soil eventually becomes a dust.

The soil structure can be improved quickly by adding certain soil additives (see p. 89). However, the soil structure will improve over time once air, water and organic matter get to work. Getting this cycle started will involve having soil tests done to find out what deficiencies the soil has and then correcting these deficiencies with fertiliser in order to get the plants started. If land is compacted it will need to be aerated with an implement that cuts into the soil without disturbing the top layer (see p. 103).

The difficulties that you will encounter when trying to improve sandy soil are that it is fragile when dry, does not hold moisture very well and nutrients are leached out of it easily. While this means that sandy soils do not tend to get waterlogged (unless the subsoil is very compacted), it does mean that they dry out too quickly and that nutrients and fertilisers tend to be leached through them. Some deep sands have no structure at all and getting a cycle of growth started can be difficult. These soils are often so well aerated that microbes burn up any organic matter very quickly.

In addition some sands suffer from a type of organic matter that is water repelling in nature (hydrophobic). This substance coats individual grains. Sand is more susceptible to this than other types of soil because the individual grains have a relatively larger surface area.

The aim with sandy soils is to increase their organic material component and to improve the soil structure as quickly as possible. This can be done by planting deep-rooted legumes and grasses.

Clay soils have their own disadvantages because they hold too much water and therefore are prone to waterlogging (see p. 91). They do, however, retain nutrients (as long as the nutrients are not washed off the top). They can be improved by adding nonwetting sands and again by increasing the organic matter by introducing plants such as legumes and deep-rooted grasses. The roots of these plants allow air and water to infiltrate into the soil. Once the soil structure is improved, water can move through the soil rather than across the top.

With all soil types the soil structure is helped by organic matter in the soil because this starts an upward spiral of events which increase microbial, fungal and plant root activity. Once correct conditions are in place the roots of plants can utilise the nutrients in the soil, sometimes from deep below the surface depending on how deep rooted the plants are.

Soil and plant tissue testing

Soil testing tells you what is and isn't present in the soil as well as telling you how fertile it is. It sets out the nature and condition of soils, such as their acidity and salinity. Soil testing kits are available from independent soil testing laboratories, fertiliser companies and some agricultural departments. Independent laboratories are the best ones to use and will usually recommend what nutrients are needed rather than a particular brand of fertiliser (which is what a fertiliser company will do).

A soil sample can be taken by you and sent to one of these laboratories for analysis. The return report will detail the level of plant nutrition available, and the soil condition. It will recommend the types and amounts of fertiliser and soil conditioner needed. It is also possible to buy kits that enable you to test some soil properties yourself. A test for pH can be carried out easily at home with a reusable kit. Soil test kits that test the nutrients in the soil can also be purchased (from nurseries and hardware stores). These vary in their reliability.

Plant tissue tests can also be carried out (by some laboratories). These can determine the calcium (Ca):phosphorus (P) ratio and other nutrients such as total nitrogen, potassium, magnesium, sodium, copper, zinc, iron, manganese cobalt, boron and sulphur.

The pH of soil is based on the soil acidity or alkalinity and is measured on a scale from 0–14. The pH affects plant growth and therefore it is important that it is correctly balanced. A pH of 0–6 is acidic, pH 7 is neutral and 8–14 is alkaline. Most plants grow best in a pH of 6–7. If the pH is outside this range then the plant tends to suffer from certain deficiencies as the pH affects what nutrients are available to them. Acid soils are especially detrimental to legumes as the low pH prevents the formation of nitrogen-fixing nodules on their roots.

Taking samples

Take samples from areas that represent the whole paddock. Don't take samples from areas such as under a fence, in or near a 'rough', near tracks, roads or buildings. You may also want to separately test specific problem areas such as spots where nothing will grow. The analysis of these can then be compared to the analysis of the 'normal' areas. Aim to take soil tests once a year or at least every three years from the same spot.

Even if you can only afford to improve one paddock each year, taking samples from your whole property will help you to decide which paddocks to start with. It is better to aim for full fertility in one paddock than to go for part-fertility in all of the paddocks.

Soil additives and fertilisers

Soil additives include materials that improve the soil condition and structure and add nutrients. Some do more than one job. It is very important that soil tests are carried out first to determine your soil's particular failing before adding anything. Some additives may only need to be applied once, others more frequently or at least until the soil is healthy and able to sustain itself.

Common soil conditioners include lime/dolomite, bentonite and gypsum. Calcium is required by acid soils that 'lock' up certain plant nutrients making them unavailable to the pasture. Lime and dolomite are mainly calcium carbonate and dolomite also contains magnesium carbonate. Both are equally good, so which one you use depends on availability unless your soil lacks magnesium (quite rare).

Bentonite is used for many applications including sealing dam walls. It is a soil conditioner and fertiliser; it neutralises acid soils, reduces ground water contamination and is non-toxic to livestock and fish. It is particularly useful for sandy soils that do not hold water as it swells when wet and holds many times its own weight in water.

Gypsum is an additive for clay soils as it improves soil texture, drainage and aeration. It is also a fertiliser and is pH neutral. Gypsum is calcium sulphate di-hydrate and it is useful in soils that need the sodium:calcium balance to be restored. Gypsum is mined in Australia and is imported into New Zealand.

Other soil conditioners include ground dressings such as poultry manure, compost, kelp and fish emulsion. These additives are both soil conditioners (because they add organic matter) and fertilisers (because they add nutrients) at the same time. Adding composted horse manure that is produced on your property is an excellent way of conditioning your soil for free. If you are buying ground dressings you need to calculate the cost of the product taking into account the nutrients that they contain. Due to their bulky nature, they can become expensive, as a large part of the cost is due to cartage.

In the case of straight fertilisers you have a choice of using chemical or organic fertilisers with the preference being organic. Fertilisers add nutrients but not organic matter. This is fine if you get the correct balance of nutrients as the plants will grow quickly and produce their own organic matter.

There is no doubt that fertiliser has a place in agriculture especially when improving and establishing new pastures. However, once the correct balance is in place it should be possible to reduce fertiliser use. Plants access nutrients from the soil that are a result of decomposition of the deeper layers. Deep-rooted plants do this better than shallow rooted plants, therefore once plants are in place this cycle takes place naturally. The manure of grazing animals adds to this process so unless the paddocks are used for cropping (hay or feed crop etc.) and that crop is then sold off the property, less fertiliser is needed as time goes on. If the crop is fed to resident animals the manure should be used as fertiliser to put back most of the nutrients. Fertilising a crop paddock or a pasture with horse manure will return much of the phosphorus and potassium consumed by horses. Nitrogen will need to be added if the pasture has few legumes.

Overfertilisation leads to excess run-off into the waterways causing all sorts of problems such as algal blooms. Soluble chemical fertilisers can kill earthworms and destroy microbiological soil life, they also have an environmental cost as non-renewable natural resources are used in the making of them.

Green manuring is a method of fertilising by growing a crop of leguminous plants that are then slashed or mowed and either left on the surface to decompose or turned back into the soil. This is best done before the plants reach maturity (once mature they have used up much of the available nitrogen and are not as effective). As the plants die, they release nitrogen into the soil which helps the next crop of plants to grow. Other benefits are that the roots open the soil allowing air and water deep inside. Crops that can be used for this purpose include certain types of peas and beans, vetches, lupins, clovers and lucerne.

Different parts of Australia and New Zealand have different deficiencies in the soils, even within a relatively small area fertiliser requirements can differ widely. The most limited plant nutrient in Australian soils is phosphorus. In New Zealand some soils are deficient in selenium or cobalt. Any fertiliser should only be applied if soil tests (or tissue tests) indicate their need. This is because some nutrients such as selenium are poisonous if present in high amounts. In addition, adding nutrients that are not needed is not only costly but causes imbalances to the soil and run-off pollution.

Contact your local agriculture department for their advice on which soil additives are necessary and available in your area.

Land degradation

Land degradation occurs from various forms of mismanagement. Some soils are much more susceptible to land degradation than others. It is a landowner's responsibility to identify possible problem areas on the property and take action before they become degraded. It is far better to prevent it happening than to have to make repairs however sometimes we have no choice as the damage has already been done.

Horse behaviour, including grazing behaviour, can contribute to land degradation. Horses, especially when healthy, young and fit and receiving supplementary feed, tend to use pasture as an exercise area. Even quiet horses walk a lot while grazing which contributes to the wearing away of the protective surface of grass. If horses are

Figure 6.2 Bare, degraded, over-stocked paddocks Credit: Jane Myers

paddocked separately they will walk fence lines creating tracks that become erosion channels. A property that has degraded land, as well as causing environmental problems, leads to adverse criticism of the horse industry. In time this can lead to an increase in legislation restricting horse ownership on certain properties (this has already happened in some areas of Perth, Australia). Therefore it is vitally important that property owners manage their land in such a way that it does not become degraded but actually improves in condition.

Waterlogged soil

Waterlogged soil is saturated soil where the spaces in the soil are full of water rather than air. The water is not draining either because the subsoil will not allow it or the area that the water usually moves on to is also saturated. If horses are allowed access to waterlogged soil their hooves compress these spaces (pugging) worsening the problem and causing land degradation.

If you have waterlogged soil on your property it will need careful management if it is not to become worse. Allowing horses to graze on waterlogged soil will lead initially to muddy conditions as the land undergoes pugging. The soil then becomes compacted. Later, when the land dries out, it becomes bare soil which leads to dust and erosion as the soil is blown or washed into waterways.

Mud has many disadvantages:

- it creates a breeding ground for insects and bacteria etc.
- it can be dangerous and cause injuries (horses and handlers can slip or fall)
- it makes caring for horses harder work
- repeatedly wet then dry conditions are not good for hooves
- feeding horses on muddy ground leads to soil/sand ingestion
- it can cause skin conditions such as mud fever, greasy heel, thrush and rain scald
- manure further compounds this problem as it holds more water than soil.

Depending on how much of your property has a tendency to become waterlogged gives you various options. Small areas (or even large areas on a larger property) can be

Figure 6.3 A muddy area Credit: Tina Wilson

fenced off and allowed to become wetland or regenerate as wetland if that is what it was before. If the waterlogged area is near a watercourse already then this area as well as the watercourse should be fenced. Trees can be planted in wet areas (once fenced off) to create habitat for wildlife. Check which trees are local to the area and which ones do well in wet soil.

If fencing a block of land from the beginning, divide the block into dry paddocks and wet paddocks so that horses can be rotated around the property. This means keeping them in the dry paddocks in the wet season and vice versa. Until you have owned the land through at least one cycle of seasons it is not always possible to know which areas do what at what time of year. Using temporary electric fences to start with can save making expensive mistakes.

If your property tends to become wet all over at certain times of the year then confine your horses to the sacrifice yards (see p. 115) until the property has dried out. Allowing horses access to waterlogged paddocks will do a lot of harm both immediately and further down the track.

Planting certain water loving trees such as melaleucas around the edge of each paddock will help to dry paddocks out. However, this takes time and the paddocks must be protected from hooves in the meantime.

If the waterlogged area is a gateway or other high traffic area this can be improved by surfacing with gravel (stones, etc.) or woodchips. This surface will need to be at least 10 cm thick, preferably more. Try to apply the surface before the mud occurs, i.e. at the dry time of the year. If possible put a membrane down first, such as old wool carpet turned upside down or a tarpaulin. Make sure that any water that runs over the area is diverted and pick up manure on a regular basis. Try to avoid having horses standing around in this area.

If you are rejuvenating wet areas that have become degraded in order to use them for grazing (at the appropriate time of year), plant grasses that have a tendency to mat as this resists the damage that hoofs cause (for example phalaris or fescue). Kikuyu is excellent as a hardwearing ground cover. Only allow horses to graze these areas at times that will cause minimal damage, at other times slash/mow or use other animals such as sheep, goats or geese to keep the grass short.

Bare/compacted soil and erosion

Typically land degradation occurs first as bare soil, then compaction, then surface erosion. It all begins with a combination of hoof and grazing pressure (horses' hooves, especially when shod, shear the grass causing it to die and disappear). Horses have the ability to graze to ground level, therefore a combination of factors including overstocking paddocks, allowing horses to stand around in areas for long periods of time (such as at the gate waiting to come in), and allowing horses to overgraze paddocks results in land degradation leading to erosion. If the land is sloping this further exacerbates the problem as water will run over the bare land more quickly causing erosion channels.

These conditions typically occur in high traffic areas such as yards, laneways, near gates around feed and water points, and shady areas, i.e. under trees. The weight of the horses compact the soil and without the cushioning effect of grass the problem is exacerbated. Compacted soil lacks oxygen as the air is forced out. At this stage even if the

Figure 6.4 Bare, compacted soil and erosion Credit: Jane Myers

horses were removed, plants can no longer grow there (apart from some very hardy weeds) due to the hard soil and lack of air spaces.

When it rains the rain cannot penetrate (due to the compaction) and the water runs off taking the topsoil with it. This is the start of erosion. These areas are also prone to further erosion when the weather dries out and they become dust bowls. The wind then takes yet more topsoil. It is easily demonstrated that a potential erosion situation becomes a downward spiral.

Tunnel erosion

Another type of erosion and far more insidious is tunnel erosion. This is equally damaging to the land, if not more so. Unless you know the signs to look for, you may not be aware of it until a lot of damage has been done.

Tunnel erosion occurs when water finds it easier to travel underground than on the surface. This happens in the case of soils that have unstable subsoil. In an area that was previously treed (most areas) water was used by the roots of trees and the subsoil rarely became saturated. After tree clearing and the introduction of rabbits, tunnels (from rotting roots and rabbit warrens) open up the opportunity for water to move underground. As with all erosion this gathers momentum and the tunnel gets larger until eventually the roof of the tunnel caves in resulting in a huge erosion channel. Usually before it gets to this stage small holes can be seen in the ground that look like rabbit holes. These are sometimes caused by horses' hooves as the weight of a horse can cause them to break through the 'roof' of a tunnel. This is a very dangerous situation for horses as serious injury including broken legs can result from holes in the ground.

This is just one case where the welfare of your horse is closely linked to the welfare of the land. Degraded land also has the following effects on horses:

- it results in less grass
- the hard ground jars legs
- the dust that is created affects lungs and health of horses, especially young horses
- a favourable environment for weeds is created which can be bad for horse health.

Land degradation should be prevented from occurring and any problems must be tackled as soon as they become apparent. Some of the strategies that can be used to prevent and control land degradation are listed below.

- Tunnel-eroded areas must be fenced off and revegetated with deep-rooted trees and plants to hold the soil together. As with other types of erosion, water should be diverted to partially halt the problem.
- Allowing horses on to paddocks when they are either too wet or too dry, such as in a prolonged dry period, leads to damage of pasture which starts the downward spiral. There comes a time when allowing horses out is doing far more damage than can be tolerated. During a drought this dry period goes on and on and you may have to keep your horses off the land for a long period of time.
- Plants are essential in the war against compaction and erosion. Their roots hold soil together, provide organic matter to the soil, take air and allow water into the soil which all help to bring in the earthworms which are both an indication of and a necessary part of healthy soil.
- A healthy vegetative cover is essential for good soil. Even weeds can be better than nothing and indeed they do many or all of the things that preferred plants do, apart from provide safe nutritious fodder for stock. Therefore weeds should only be removed when you are ready and able to replace them with a preferred species. In fact out-competing the weeds with the preferred species is better than complete removal in many cases, if it ensures that the land is never completely bare. Planting hard-wearing grasses such as kikuyu in high traffic areas works well.
- Removal of horses is important for the repair of degraded land. This can either be done by fencing off areas that are to be rejuvenated or removing the horses altogether. Usually a balance of the two works best. Areas that are prone to erosion may have to be fenced off from horses. This includes steep hillsides and creeks etc.
- Reduce the loafing time that horse spend in the paddock each day by removing them for at least part of each day. There is a period (or several) during a day that horses spend just loafing (see p. 9).
  By controlling the amount of time the horses spend in the paddock this behaviour can be manipulated so that the horses graze when in the paddock and loaf when in the sacrifice yards (see p. 115). Turning the horses out for two grazing periods per day is one way of ensuring that when horses are in the paddock they get on with eating. The rest of the time the paddock gets to rest and recuperate. This practice, called 'limited grazing', should be carried out in conjunction with other grazing systems (see p. 111).
- Only have horses shod if it is really necessary. Many horse owners shoe horses out of habit rather than a real need (see p. 29).
- Shelterbelts protect the soil from the wind (wind carries your exposed soil away). Trees and other vegetation also help to bind the soil (see p. 175).
- Control and prevent water from running across areas of bare soil. You need to understand how your land drains, where the water runs and where it pools.

Next time it rains heavily go outside and watch where the water is going. Areas that have a lot of run-off must be protected with vegetative cover. Water should be directed across a slope rather than allowed to run down it. Fencing on contour lines, improving the ability of the land to absorb water and installing drainage ditches to divert water help too. Also logs, tree branches, bales of hay/straw and tyres can be used to divert water either permanently or until an area has revegetated.

- Do not overstock your land and do not allow overgrazing (move horses on when the grass is down to an average of 5–8 cm). This will give plants a chance to recuperate and regenerate. Allow the grass to reach an average of 15–20 cm again before letting horses have access to it again.
- In very small paddocks that are difficult to keep vegetated, consider amalgamating with other small paddocks and renovating the area. If this is not possible because you do not have anywhere else to put the horses, then the areas may require surfacing if grass will not grow there anyway. These areas then become sacrifice areas (see p. 115).
- Cover any bare patches in the paddock, use either composted manure, leaves from under trees, old hay or grass.

Salinity

There are two causes of salinity – irrigation and land clearing. Irrigation salinity is caused by salt in irrigation water being deposited in the soil and accumulating. Dryland salinity is a result of wholesale clearing of trees. When trees (deep-rooted plants) are removed and replaced with shallow-rooted crops the ground water rises to the surface of the soil. As it does, it brings salt that is meant to be underground to the surface. This salt may then also be washed into waterways causing problems beyond the immediate area. These salt deposits on the land cause plants that are not salt tolerant to die and plants that are salt tolerant to invade. Also, rising water tables cause waterlogging, which in turn causes plants that cannot cope with wet roots to die.

Once affected, land that has salinity is difficult to manage. Remedies include planting trees, planting salt-tolerant bushes and grasses, not using the land when wet or even permanently fencing off areas and revegetating with trees and bushes that are native to the area. Your local Landcare officer or Agriculture Department will be able to give you invaluable advice regarding your locality.

Suitable pasture species

Different pasture species have different values in horse paddocks. The species you have in your paddocks (or want in your paddocks) vary due to the local climate and soil conditions. There are lots of available grasses and it is impossible to list them all here. However, it is relatively easy to find out what you can grow in your paddocks. Start by speaking to a local seed supplier.

The kind of pasture that you want/need will be determined by its purpose. Does it have to be a tough species that the horses can run around on and that resists erosion and weed invasion, or a nutritious pasture for feed? Some species can do both to some extent.

Pasture can be made up of either one species type or many species living together. The latter is better as biodiversity is far superior to monocultures. In nature there is always a variety of species in any ecosystem in order to be sustainable. Monocultures are prone to disease and pest invasion. A variety of species will give a good balance of nutrients to the horse and will withstand grazing pressure better as they peak at different times in the season. It will take time to establish a varied pasture but eventually a blend of grasses, legumes, medics and herbs should be aimed for. Horses need variety in their diet and thrive on it.

Perennials and annuals

Perennials are plants that, once established, will keep growing year after year. Annuals are plants that grow, set seed and die in one year. Establishing perennials in your paddocks is important if you want to have permanent pastures that are reasonably easy to take care of.

Legumes

Legumes, as well as being nutritious for horses, play a very special role in the paddock. Legumes have the ability to take nitrogen from the air and 'fix' it. They do this by forming an association with a bacterium which draws on nitrogen in the air. These special bacteria, called rhizobia, live in lumps (nodules) on the root of the legume. They themselves grow by using carbohydrates that are made by the plant. This is therefore a symbiotic relationship that benefits both species as the nitrogen that the rhizobia release is used by the plant. Also when the plant dies or is cut back (by slashing or grazing) this nitrogen is released into the soil making it available for other plants. Legumes that are useful in a pasture include clovers, lucerne and medics.

In paddocks that show up (from a soil test) as being low in nitrogen, consider sowing legumes such as lucerne and clovers. This can either be as a crop in the year before the grasses are sown or as part of a pasture mix (e.g. lucerne can be added to the pasture sowing mix at the rate of about 0.5–1 kg per hectare). Sowing a legume such as lucerne has many benefits. By repeatedly slashing or grazing the plant, the roots die and then regenerate releasing nitrogen into the soil. At the same time the dead roots create organic matter in the soil and allow air and water into the soil. The plant also spreads and thickens and as time goes on horses can be allowed longer periods of grazing.

Figure 6.5 Legume Credit: Jane Myers

In general, lucerne and clover contain more calcium than the grasses and are very useful when sown with tropical grasses such as buffel grass, kikuyu, and panic grasses. These kinds of grasses have high levels of oxalates which can cause a calcium imbalance in horses that graze exclusively on these species (see p. 98).

Palatability

Palatability in this book means what the horses prefer to eat. In a mixed pasture, therefore, some species may be eaten out while others are left. If horses are left to their own devices the paddock will eventually only contain the species that they do not like as they overgraze the species that they do. Good grazing management will greatly reduce this effect as the horses are not as able to choose and graze out favoured species especially with strip/block grazing. Most plant species become less palatable as they become more mature (see p. 114).

Drought tolerance

In much of Australia and parts of New Zealand drought is a regular phenomenon and occurs more often in some regions than others. The Australian native grasses (wallaby grass, kangaroo grass, snow grass and spear grass) cope well with drought and frost. They respond well to light fertilisation and the presence of legumes. However, they do have lower yields than many other grasses. Speak to your local Agriculture Department to find out which ones are likely to grow in your area.

A period of drought will usually result in poor regeneration of annual grasses due to a reduced seed set. Some weeds do very well from drought such as fireweed, ragwort, capeweed, dock and thistles. There is also more chance that these weeds will be brought on to the property in hay that will be purchased in a drought.

Perennial species will undergo significant reductions in plant numbers during a drought with perennial ryegrass being the least tolerant, then cocksfoot/tall fescue and phalaris. Because phalaris is dormant over summer it copes better than the other perennials. Lucerne, due to a large taproot, can survive drought reasonably well as long as it has breaks from grazing pressure.

Strength and persistence

Some grasses have more strength and persistence than others. These grasses are good for areas that get a lot of traffic such as gateways and laneways as well as the whole paddock. Kikuyu can be planted from runners or seeds; it is best sown with white clover and lucerne. This is a good species for horses if the legumes are maintained even though it has a lower nutritional value than some other grasses.

Rhodes is a grass species that is a perennial with deep roots. It is also salt tolerant. It produces runners or stolons that cover the ground and protect it from erosion and other land degradation. Ryegrass is good for heavy grazing as it remains longer in the vegetative state.

Nutritive value

Some species have higher nutritive value than others, such as the legumes (clovers and lucerne), which are higher in protein than most grasses.

When you only have a small amount of grazing available, it is usually better if it is high in nutritive value because the horses will have limited grazing time on the pasture. This must be weighed up with factors such as the species' hardwearing qualities because the pasture gets more pressure on smaller properties.

Climate limitations

The situation of your property will determine what kind of pasture can be grown. You need to find out if you live in a temperate, subtropical or tropical climate. Another consideration is rainfall as this varies from region to region and even within a region.

Your local Agriculture Department office can help recommend suitable pasture species for your area. Also speak with local seed retailers and other experts.

Nutritional problems with pasture

There are certain problems that can occur when horses graze pasture. You need to find out what potential problems there may be with the type of pasture that you have. It is recommended that you learn as much as possible about any likely health conditions so that you will be able to recognise them if they occur.

Ergot of paspalum

Paspalum is a tufted grass that forms a dense mat under grazing conditions. The seed head consists of two to five very thin stalks. A toxic ergot fungus invades the seed head, replacing the seed with a mass of sticky, black material, which eventually becomes hard and seed-like. Horses can take a liking to infected paspalum heads. Most outbreaks occur in late summer–early autumn, when paspalum flowers. Paspalum should be kept relatively short to reduce the likelihood of its flowering.

Ryegrass toxicity/ryegrass staggers

This is caused by ryegrass affected by a toxin-producing bacterium. It also affects sheep and cattle. It is present in grass and in hay with ryegrass in it. The symptoms in horses are recognised as a wide stance and a wobbly gait that may lead to temporary paralysis of the hindquarters. Some horses become recumbent and have to be destroyed. Affected horses should be moved quietly from the paddock and a vet should be called.

Stringhalt

The signs of stringhalt basically involve an exaggerated lifting of each hind leg as the horse moves forwards or backwards.

The high stepping hind limb is caused by degeneration in the long nerves of the horse's hind limb. This is the reason for the muscle wastage and the abnormal movement of the leg. However, the nerves and muscles can regenerate if given sufficient time and, unlike many other diseases of the nervous system, horses can recover from stringhalt.

The cause of stringhalt is still unclear. The weed flatweed (*hypochacris radicata*) (cats ear/false dandelion) is commonly found in paddocks in which stringhalt occurs. It is likely that there are several possible causes rather than one direct cause. If you suspect that your horse has stringhalt remove it from the paddock and call a vet.

Tropical pastures and 'big head'

Big head is a calcium deficiency in horses caused by grazing introduced tropical pasture grasses. It is caused by crystals of calcium oxalate in the grass blades that prevent the horse from absorbing calcium from the grass during digestion. Cattle and sheep are not affected as their rumen bacteria break down oxalates and release the calcium for absorption.

The signs of big head include:

- swelling of the jaw as the bones enlarge
- lameness – horses appear stiff and have a shortened gait
- ill thrift – loss of condition on pastures that look nutritious.

The hazard is greatest when introduced tropical species such as buffel grass, green panic, setaria, kikuyu, guinea grass, para grass, pangola grass, signal grass and purple pigeon grass provide all, or almost all, of the feed available. Australian native grasses have not caused the disease, nor have other grasses such as rye grass, Rhodes grass, the paspalums, the couches and creeping blue grass.

If only hazardous grasses are available, encourage the growth of a legume component in the pasture to provide a source of feed free of oxalate, and feed a calcium and phosphorus supplement.

Management practices

Weeds

What is a weed? A plant that is in the wrong place at the wrong time is one way of defining a weed. The term 'weed' covers a huge range of plants and what is a weed to one person may be a beneficial herb to another.

Weeds may be declared poisonous plants which are dangerous to stock and the environment and the economy (see p. 180). These types of weeds usually have vigorous characteristics which are why they are so important. A good example is ragwort which is a problem in some parts of Australia and in New Zealand. Other weeds, rather than being dangerous, are plants with simply very little or no feed value. They reduce pasture yield by taking up space and compete with pasture grasses for moisture, sunlight and nutrients. These weeds are still a problem as they rob you of feed and cost time and often money to control. They may also be highly invasive. Some examples of such weeds are blackberry, gorse (furze) and thistles.

Weeds can be good indicators of soil conditions. So much so, that eradicating some weeds is simply a matter of changing the soil conditions in one way or another.

Weeds can also have the benefit of opening up the soil so that more desirable plants can grow. It may be possible to out-compete these weeds with more favoured plants.

As a landowner you are expected to control weeds on your property (you may even have a legal obligation in the case of some noxious weeds) and sometimes those on the area between your property and the road (the nature strip). In most areas there are 'statutory' requirements to ensure that weeds do not spread, threaten the wellbeing of landowners, the community in general and the health of livestock. Check with your local Agriculture Department about the regulations that will affect you and your property.

Weeds arrive on your property in various ways. They blow in from neighbouring properties and the roadside; they are carried in by birds (in their droppings); they come in with loads of soil and gravel; they stick to other objects such as car tyres and animals (burrs do this); they grow creepers (underground and above ground) and invade from neighbouring properties; they come in with new or visiting horses' droppings and they come in hay and other fodder that is brought onto the property.

Properties that have not been managed correctly in the past may have large weed infestations that are difficult to get rid of, especially as they will often be part of other land degradation problems such as compacted soil. Poor paddock management allows weeds to take over. For example if horses are allowed to overgraze areas they then become bare and weeds start to take over until eventually there are more weeds than grass. Weeds invade areas that have no grass because they exploit ecological niches. This happens especially in places such as water points and feeding areas.

Knowing your weeds will enable you to exploit their weaknesses. The library will have books and your local Agriculture Department will have fact sheets that enable you to identify weeds on your property and provide information on how to deal with them. Also see if your local council has a Weeds Officer. If so, this person will be very useful in helping you to deal with weeds. (See Chapter 12 for resources on organic weed control and plant identification.)

Control of weeds

There are various strategies for dealing with weeds ranging from minimum to maximum intervention. Often a combination of more than one strategy works best. With many weeds it is very important that they are not allowed to set seed as this means that there will be even more of them next year.

Strategies for weed control are described below.

Mechanical control

This can give good results but must be done carefully so as not to cause more problems such as spreading seed.

- Mowing can reduce the seed setting of some weeds but must be done at the right time or you may be spreading seeds. Also it is a good strategy for tall plants but may not be effective for low growing plants
- Some weeds can be thrashed by hand with a whipper-snipper or a scythe, for example nettles are actually good for horses but they will only eat them when they are wilting.
- Hand pulling/hoeing by hand. This is very hard work but is fine for small infestations and for keeping on top of paddocks or areas that have been treated more aggressively. It must be done regularly to be effective. Rather than just pulling weeds out always replace with seeds from a preferred species.

Organic control

Organic control often takes longer than chemical control but is better for the environment.

- Flame and steam can be used for weeding. Both methods are ecologically sound.
- Cross-grazing with other animals can be useful, especially sheep and goats which are more resistant to certain toxins such as the poisons in fireweed (from the same family as ragwort). These animals are also invaluable for controlling blackberry, gorse (furze) and Paterson's Curse which is a huge problem in most of the southern areas of Australia.

- Weeds can sometimes be out-competed with other more desirable species.
- Mulching can be used to smother weeds in areas that you plan to replant later or that need the protection of mulch permanently (such as around trees and pathways/laneways). A huge variety of materials can be used as mulch, such as carpet, newspaper/cardboard, woodchips, or stone chips/gravel.
- Paddocks can be intensively grazed prior to planting new pasture. This will depend on the species present. This should be done in the summer/dry period before you plan to plant the new pasture.
- Changing the chemical balance of the soil is an effective way of controlling certain weeds.
- Some weeds can be controlled biologically by using the weed's natural enemies such as specific weevils for Paterson's Curse. Speak to your local expert, as each area is different.

Chemical control

Chemical control can give fast results but can be harmful for the environment.

- Certain chemicals will kill every plant in the paddock. Others are more selective, however, even then, they may often kill the desirable along with the undesirable for example a broadleaf spray will kill clovers along with weeds.
- When weeds are killed in this way the paddock is left bare until the grass grows again so it needs to be done at the right time, i.e. not at the beginning of a hot dry summer.
- Plants must only be sprayed when growing, when green and prior to setting seed. This will give you the best result. Spray on still days when at least 12 hours of dry weather is expected. Follow-up spraying will need to be done some weeks later to get the ones that were missed the first time.
- Do not spray within 100 m of dams, wetlands, creeks etc. unless using products that are safe for aquatic life.
- If using a contractor, make sure they are experienced and have the right equipment and operating certificates. Your local Agriculture Department should be able to recommend contractors.
- Animals will need to be removed from the pasture or area that is sprayed for the specified time (read the instructions). After spraying the weeds can be more palatable to animals for a period of time which can result in poisoning, depending on the type of weed.

Preventative measures

There are also strategies you can put in place to ensure fewer weeds can establish themselves on your property.

- Being very careful when purchasing hay is one way of controlling weeds. By buying directly from the grower you can find out what weeds are a problem in that area and check to see if that property is infested. Check the hay before buying (preferably before baling) by asking to see a bale opened. The farmer will probably only be willing to do this if you are about to buy a reasonable amount. If you are buying from a feed store or hay merchant, ask if they have checked

the quality of the hay and whether they are prepared to guarantee the quality. With all purchased hay (as opposed to home grown) it is advisable to feed it in yards or stables rather than in the paddock. You will be able to monitor and control any weeds that crop up in these areas far easier than if they get into your paddocks (although doing this does not guarantee the weeds not getting into the paddocks, it just reduces the risk).
- Don't allow your land to become degraded and cover all bare areas in paddocks, laneways and yards.
- If anyone visits your property with a horse pick up the manure and compost it. Also beware of the hay that they are using.
- Establish multi-storey shelterbelts around your property to reduce the number of weed seeds blowing in from neighbours and the roadside.
- Make sure pasture seed is certified as weed free (beware of buying cheap seeds).
- Experiments have shown that weed seeds pass through the digestive system of horses relatively unscathed compared to other animals where they do not survive (ducks, chickens, cattle and sheep). By keeping chickens and ducks and allowing them to free range and 'sort' through manure for seeds helps to vastly reduce the number of seeds that get into the ground.

Drought

Caring for pasture in times of drought involves reducing the grazing pressure as much as possible during the drought itself and for some time afterwards in order to allow pastures to recover. During the drought grazing animals will have to be fed on hard feed and/or hay. The horses will need to be confined to yards or stables once all of the paddocks have been grazed down to an average of 5–8 cm to minimise land degradation. Feeding horses during a drought sometime involves having to buy hay from sources that you would normally avoid, due to the risk of weeds, therefore feeding in yards becomes even more important.

Once the drought is over resist the temptation to turn the horses out until the pasture has fully recovered and start with just a few hours at a time. See Chapter 12 for resources on drought management of horses.

Drainage

Good drainage is essential for plant health as a wet, poorly drained paddock will have reduced pasture growth and the movement of horses around the paddock will destroy pasture and create bare areas. During wet periods horses must be removed from waterlogged paddocks (see p. 91).

Paddocks should not be allowed to stand underwater for long periods as this will kill many pasture grasses. This happens in many horse paddocks simply due to compaction of the soil. First and foremost, the soil condition should be improved so that it can take in water and use it to grow plants. Certain plants such as lucerne and deep-rooted grasses (such as Rhodes grass) take water (and air) deep into the soil by opening up the soil with their long roots. They also create organic matter in the soil as their roots grow and die on a continuous basis.

Figure 6.6 A French drain Credit: Jane Myers

Young plants cannot get through hard compacted soil. In this case the soil should be opened up first with an implement that cuts into the soil without any turning or mixing of layers. Various implements can do this such as a subsoil ripper, a Yeomans/Keyline plough or a scarifier. It is very important that this is done along the contour lines of a paddock rather than across them so that rainwater enters the soil rather than runs downhill creating erosion. This may need to be done several times over a period of time in order to maximise the absorption rate of the land.

Certain soil additives can also improve the absorption rate of the soil (see p. 89). Always check with an expert before applying soil additives otherwise you may actually make the problem worse.

Water that cannot be absorbed should be channelled, by constructing drains, so that it runs towards waterways and any dams. Aim to channel water along contours whenever possible, gradually taking water down to the water level rather than allowing it to rush downhill. This can be difficult as land has usually been subdivided without taking contours into consideration (both within a property and between properties).

Water can be channelled using materials placed on the surface of the land such as logs, tyres, bales of hay/straw or mounded earth etc. or by creating drainage channels with machinery. It may be possible for you to hire machinery or to hire a contractor to make them. The drains can be in the paddock but it is better to have then along fence lines and laneways. Drains can be open and lined with gravel or grass, they can be loosely filled with gravel or rocks (a French drain), or they can be underground for example 'aggi' pipe in channels of gravel.

Slashing/mowing

If you are not planning to use other animal species to eat down the grasses that horses leave, then the paddock should be slashed or mowed immediately after the horses have been moved on to the next paddock. This procedure is also sometimes called pasture topping. The grass should be cut to a height of 5–8 cm level with the rest of the paddock.

The difference between slashing and mowing is that slashing simply cuts the grass once, leaving long stems lying on the ground, while mowing chops the grass (mulches it) into small pieces. The latter is usually preferable as the grass then breaks down sooner adding condition to the soil. Tractors can be fitted with slashers or mowers; domestic lawn mowers and ride-on mowers mulch.

Slashing or mowing helps to remove old, tough, dry grass and encourages the growth of young green grass. This results in more uniform regrowth. Mowing grasses before they set seed will encourage them to produce leafier and more nutritious vegetation. Young immature plants have more leaves than stems and are more nutritious than older stemmy grasses. This is because there are two to three times more nutrients in the leaves than in the stems of plants.

Mowing the pasture during a period of rapid pasture growth can also be advantageous. Ideally the pasture should be mowed to a height of 8–10 cm. This keeps the pasture at a similar level over the whole paddock, encouraging the horse to graze the whole area.

Mowing weeds before they set seed is an important weed control strategy for certain upright weeds such as thistles. However, broadleaf weeds that are low to the ground and grow outward rather than upward may not be as effectively controlled by mowing. Mowing weeds at the wrong time (when they have seed heads present) can actually spread them around your paddock.

Harrowing

Harrowing is a manure management strategy which has the added benefit of improving pasture. By spreading manure around the paddock rather than allowing it to stay in areas called 'roughs', the paddock will be more evenly grazed in the future and will benefit from the nutrients in the manure (see p. 193).

The best time to harrow a paddock is after it has been grazed and the animals have been moved on to the next paddock unless this coincides with a period of heavy rain. Heavy rain will wash harrowed manure off the paddock much more easily than undisturbed manure. It is preferable to slash any long grass that has been left in the paddock before it is harrowed and the harrows will then distribute this around the paddock at the same time. If a hot dry period or a period of frost follows harrowing it will kill most of the parasitic worm larvae in the manure dung pats as they are exposed to the air. If the weather is warm and humid it does not kill the larvae as effectively and the paddock should be rested until hot/dry or frosty weather occurs before putting horses back on it. Either way the paddock should be rested until the grass has reached an average height of 15–20 cm.

Another positive effect of harrowing is that it eventually eradicates the grazing behaviour of horses that causes roughs and lawns (see p. 7) because the manure ends up spread evenly around the paddock. This is a controversial subject. Some people believe that this will result in the horses rejecting the whole of the paddock, rather than just the roughs, and that only roughs should be harrowed so that manure is not spread into the lawns. The problem with doing this is that these areas become lost to

Figure 6.7 Pasture harrows Credit: Jane Myers

the available grazing area, as they tend to get larger over time. Also this compounds the nutrients in these areas with the lawns becoming increasingly depleted (unless they receive supplementary fertilisation). In practice, the rejection of the whole paddock does not happen and as long as you employ other methods of worm control, such as regular worming, resting paddocks after harrowing, and harrowing at the right time, the horse will not become overburdened with worms and the paddocks will end up with even growth.

Coping with too much pasture

Sometimes you may find yourself in the enviable position of having too much pasture. This will probably be only at certain times of the year but will still need to be dealt with.

Long grass is a fire hazard. However, if it is well away from the house with fire breaks in between, or the surplus occurs at a low fire risk time of year, it does no harm to allow the grass to grow tall and set seed. In fact the grass protects the soil so it is doing a valuable job. Even though this grass will be lower in energy it can be grazed later in the year when the other paddocks have stopped growing. For horses that tend to get overweight easily this may be the only time of year when they can graze for long periods. This area will need to be slashed if it is not evenly grazed before the next growing season so that the new grasses can grow.

Another option is to use other grazing animals such as cattle and/or sheep (see p. 112). These animals can be shared or borrowed if necessary.

If there is nothing to eat the grass and the long grass is a fire hazard then you will need to slash it or mow it.

Conserving pasture (hay and silage)

On some properties there is so much grass at certain times of the year that making hay or silage (see p. 21) can be considered. As with many things, there are advantages and disadvantages to doing this, as shown below.

Advantages

- It saves on the costs of buying hay or silage.
- If hay is kept dry (in a shed) it can be stored for several seasons, although it will lose feed value over time. Well-packaged silage can be stored indefinitely without losing quality and does not require a shed. This can help to offset the effects of a future drought.
- It solves the problem of what to do with all the extra grass.
- It reduces the incidence of weeds invading your property through buying in feed.

Disadvantages

- It is not cost effective for a small property to buy the necessary hay-baling/silage-making machinery, plus it takes time to make hay. The hay may need you when you are otherwise busy.
- Contractors are more cost effective but are in high demand at the time that you need them (everyone's hay or silage tends to be ready to be cut and baled at the same time). What you can make will be dependent on the availability of contractors in your area.
- Taking forage off the land removes large amounts of nutrients, which must be put back either in the form of bought-in fertiliser or composted manure gained from the animals that ate the forage.
- There is the chance that the weather can turn and the crop is spoiled thus wasting money spent on contractors etc. This risk is reduced if making silage rather than hay.
- Unless your pastures are very good then you will only end up with poor quality hay or silage. In this case it is better to spend the money on good quality forage from a reputable source. This will add to the wellbeing of your animals and your land (through manure and grass seeds).

A crop can be sown especially for making hay (or silage), for example oats to make oaten hay. Oats will reduce weeds and will utilise the available legume nitrogen. A crop that is sown especially for hay should be able to ensure a high quality yield (as long as the other factors such as rainfall and fertilisation are present).

Improving pasture

The effects of pasture decline are many and affect you, your horse and the environment. Pasture decline means that horses need extra feed (therefore extra expense), your horses' health declines from a lack of fresh grass grazing, from dust and from eating weeds. Horses may also start to ringbark trees, be more aggressive with one another and eat soil. The environment also suffers from land degradation problems.

Unless your pasture is already really good you should plan to improve it. This will pay off many times over with an increase in the grass available and better ground cover. Signs that your paddocks require renovation are bare patches, weeds and the presence of roughs and lawns.

You need to plan prior to renovating a paddock. Without planning it may go wrong and you will be reluctant to try again. For example establishing new pasture will be much more successful in years that are not in drought and during the highest rainfall time of the year. The time of year is critical as there must be enough water in the soil. Also plan how you will manage the rest of the property because newly renovated pasture will not be able to be grazed for some time, putting more pressure on other paddocks. Do a lot of research beforehand about what plants to re-sow. Weed control is top of the list and should start long before renovation to reduce the seed bank of the weeds.

Before you start, consider the strategies listed below.

- Have soil tests done to help you determine which area will be renovated first. It is usually better to pick an area that has the most chance of success to begin with.
- Work out your budget, including buying in extra feed because you will be a paddock short for some time.
- Get expert advice.
- Control weeds.
- Fertilise if necessary.
- Select your species and source quality certified seed.
- Fence off the areas to be renovated; this can be done with portable electric fencing.
- Find out if and when contractors are available if needed.
- Prepare the ground for sowing.

There are various levels of pasture improvement and sometimes different paddocks may need different levels.

A paddock can be enhanced by simply improving current management practices such as mowing, harrowing, soil testing (and using this information to consider fertilising), controlling weeds, covering bare patches, improving grazing management strategies (i.e. strip/block, rotation, cross-grazing and or limited grazing) and collecting manure on small paddocks.

In addition to the above methods, a small amount of cultivation (i.e. scraping the soil with harrows in order to break the surface of the soil) can be done. Seeds can be broadcast (spread across the surface) by hand or seed spreader and either left on top of the soil or harrowed in.

Figure 6.8 Ploughed land Credit: Jane Myers

a

b

Figure 6.9 (a) Pasture before renovation and (b) four weeks after renovation (due to good rainfall immediately after seeding). Chicken manure and lime/dolorite were spread on the surface and the paddock was then ripped. Seeds (Rhodes grass plus other grasses and legumes) were spread by hand. Credit: Jane Myers

A higher level of renovation is to direct drill the seed. This is where the seed is drilled into the soil, usually along with a fertiliser. With this method the old pasture can be sprayed or grazed hard before seeding and for two weeks afterwards to give the new plants a better chance. If the old pasture is left in place this procedure is called undersowing. Any established pasture weeds need to be controlled before and after seeding.

If the soil is compacted then the paddock can be ploughed or ripped (see p. 103) and then reseeded. New seed can be broadcast by hand or with a seed spreader. A seed spreader is not usually necessary unless you are seeding many acres at a time. Spreading seeds by hand in a two acre paddock takes about one hour. The old pasture may be either killed by spraying before seeding or left alone. This is the most expensive solution and can fail if it is not carefully planned and carried out. There is a good chance that weeds will get a hold and it will be some time before horses can be allowed back on the paddock due to the looseness and fragility of the soil. Grasses with deeper roots can be grazed sooner than shallow rooted plants.

Deep cultivation shatters organic matter, dries out the soil, destroys bacteria that is necessary for plant growth and exposes soil to the air. This 'burns' off the organic matter in the soil. Only deep soils should be ploughed (if at all) because ploughing shallow soils takes the thin layer of topsoil underground. Ripping is better because it loosens the soil without turning it or mixing the layers. Ripping rather than ploughing should always be carried out unless the top soil is deep.

When the new pasture begins to grow it should be maintained in the elongation phase of growth (see p. 84). This will encourage the plants to spread and thicken which will result in the ground being covered as soon as possible. Initially this can be done by mowing or slashing. A ride-on mower will compact the soil less than a tractor. Horses can be allowed short periods of grazing when it is possible for them to take bites without the plant being lifted out by the roots.

Careful management in these initial stages is very important if the paddock is to grow well.

Grazing management

Even if grass is irrigated and fertilised it will not be able to cope with continuous pressure from the hooves of horses and their ability to eat grass right to the ground. It needs time to recover and time to set seed especially if it also has to compete with weeds. With correct management most of the negative effects that horses can have on pasture can be reversed. By utilising grazing systems horses can be persuaded to eat evenly and, by using manure management strategies such as harrowing, the effects of their dunging behaviour can be reduced also. Correct management also results in an increase of pasture and a decrease of parasitic worms.

It is necessary to make maximum productive use of your pasture, even if there is only a small area available, in order to protect the land, lower feed costs and give your horses a better quality of life.

In addition to employing good grazing systems, good horse management can further reduce pasture decline and land degradation and increase horse health and safety. If horses are to be confined to yards or stables for part of each day there are many ways in which you can modify your horse management to fit in with the available pasture. Some examples are listed below.

- If horses are fed hard feed before they are turned out they are more likely to run around and damage pasture. Turning them out after eating only hay for a while will make them get down to grazing sooner.
- Turning horses out twice a day for two shorter sessions rather than for one long session will result in them concentrating on grazing rather than running around or loafing.
- Turning horses out overnight, rather than through the day in hot weather, will be more comfortable for them as there are fewer insects around after dark.
- Exercising horses before they are turned out will also decrease the amount of time that they run around. Young/fit horses, especially after a period of confinement, will need to be worked or exercised before turning out. Controlled exercise is far safer than galloping around a paddock.
- Introduce horses to new pasture gradually as they will tend to gorge new pasture with sometimes disastrous results to themselves.

- Turn horses out together (with no back shoes or horse boots on the back feet) so that they graze rather than walk up and down the fence line (creating compaction and risking fence injuries) trying to get to another horse. As well as having psychological benefits for horses this is also important in terms of paddock management. If you have one horse per paddock you are likely to not have enough land to utilise good grazing systems (rotation etc.).

Grazing systems

Different grazing systems can be used to maximise pasture production and extend the grazing season. Thinking in terms of aiming to produce as much grass as possible and using your horses to help you do this will give good results. In the long term your horses will benefit by having more grass available to them and increased turnout times. Land degradation will be reduced due to maximising the potential of the land without depleting it.

Horses can be kept using different management systems ranging from 100 per cent confinement (stable/yards) to 100 per cent grazing. It is not desirable for a horse to have no access to grazing, as grazing is so important for the wellbeing of a horse. On most small properties there is a limit to the amount of available pasture and a combination of grazing and confinement will need to be used. Using grazing systems combined with confinement will increase the productivity of the pasture and will allow more flexibility in the number of horses that can be kept on a particular piece of land. With the exception of set stocking, which is not recommended, the grazing systems outlined here are all variations on the same theme of restricting horses to one part of the property while the other parts get to rest and recuperate.

With the use of good safe confinement areas, time spent grazing can be increased when pasture is available and decreased when it is not. Supplementary feed is used to make up the shortfall in pasture. As a pasture is improved over time, the hours spent grazing can be increased.

The amount of time that a pasture can be grazed without damage will also vary throughout the year and from year to year depending on climatic changes such as drought.

Set stocking

Set stocking is the practice commonly used on poorly managed horse properties where horses are allowed access to all the land all the time (either individually, i.e. one horse per paddock, or as a group, i.e. the horses have access to the whole property all the time). This practice leads to unhealthy land and unhealthy horses as the land becomes degraded. A paddock that is set stocked can still be harrowed and mowed so that some of the effects of the grazing behaviour of horses are reduced. However, this will vastly increase the worm burden of the horses as a paddock should be rested after harrowing so that worm larvae are not able to go to the next stage. Set stocking is to be avoided as a management practice.

Rotational grazing

Having several smaller paddocks rather than one large paddock allows paddock rotation, which improves pasture growth and parasite control and reduces land degradation. This method of management will help to prevent the under/over-grazing pattern present in so many horse paddocks. Paddocks should be divided according to the guidelines on page 65.

Paddock rotation allows grass species to recover where they would otherwise die out if submitted to constant grazing pressure. Horses tend to eat only what they like and leave the other species. This results in certain species, including weeds, taking over the pasture.

Horses should be allowed to begin grazing a paddock when it has reached an average height of approximately 15–20 cm. When they have grazed the paddock to an average height of 5–8 cm they should be moved to another paddock. Any areas that have less than 70 per cent ground cover or are bare, dusty, or boggy should be temporarily fenced off with electric tape when the horses have access to the paddock.

A simple way of measuring your grass is to use a ruler! Randomly test several areas by holding the ruler against the grass, do not stretch or extend the leaves of the grass. Average the recordings to get the pasture length. After a while you will be able to do this by eye.

When the animals are moved on, the now empty paddock is harrowed, mowed to an even length and then rested and allowed to regrow. At this point the horses can graze the paddock again. The length of time that it takes the paddock to recover to an acceptable grazing length depends on factors such as the time of the year and the pasture species. If the situation occurs where none of the paddocks are recovered enough for grazing then the horses should be confined to the sacrifice yards until they are (see p. 115).

On larger properties rotation can be over a long period of time, i.e. each paddock is used for a year and rested, cropped and used for other animals for a couple of years. This method is good for controlling parasitic worms. However, it requires lots of land which of course is not available on a small property. For a small property, a paddock may need to be used for several grazing periods within the year.

Limited grazing

This is the practice of removing horses from the pasture for part of each day in order to either conserve the pasture or to limit (manipulate) the amount of feed the horse consumes. This should be carried out in conjunction with other systems such as rotation, strip grazing and so on, as the paddocks will still require a period of weeks or months with no grazing pressure and for paddock management such as harrowing to be carried out.

Limited grazing is a good strategy for making your available pasture last as long as possible and for reducing land degradation. The horses must spend at least four hours and maybe as many as twelve hours (with good pasture) away from the paddock in order for conservation of pasture or reduced feed intake to be effective. This is because horses will simply condense all of their eating time into the one long session if necessary (see p. 6). This said, removing horses for a few hours each day, while not reducing their total daily intake, will reduce the amount of time spent loafing or sleeping in the paddock which will reduce land degradation. Horses cause just as much damage to the land during these behaviours as when they are grazing. All horses (even those with weight problems) must be given forage when confined if the period is more than four hours. Horses should not have long periods without forage passing through the gut as their digestive system is not designed to cope with being empty. Another alternative is to let the horses graze for two shorter periods per day rather than one long one so that their grass intake is spread over the day.

Cross-grazing

Rotational grazing using other animals has many advantages because they tend to complement each other in their grazing behaviours. For example each species will eat around the dung of other species but not their own. This is thought to be a parasite prevention strategy because most parasites (worms) are host specific, which means they can only complete their life cycle in one species of animal, so grazing animals avoid their own dung areas but not those of other species. Another bonus of using cross-grazing is that the land ends up with different kinds of manure on it.

Using other animals has many benefits. For example sheep eat woodier plants that horses leave; they eat weeds that are harmful to horses; they leave manure on the paddock; they take up some horse worms; they provide company; they can also provide wool, meat, leather and milk if desired. There is no machine available that can come close to doing all of this. Of course not everyone wants to get involved in milking their sheep, but this is a good example of how versatile animals can be.

The disadvantages of cross-grazing are that the extra animals eat the available feed and there are extra expenses involved such as worming and foot care. Horses tend to be dominant over other grazing animals including cows, and smaller species do require an area that they can retreat to in small paddocks.

In order for this system to work, the property needs to be producing grass at its optimum level otherwise the extra animals are just more mouths to feed when the grass runs out. For this reason it is a strategy that could be employed later on when the property has been improved.

The most common animals used for cross-grazing on horse properties are cows and sheep. However, any other grazing animal can be used including goats, lamas and alpacas, emus and ostriches and different kinds of fowl such as ducks, geese and chickens. Even small numbers of kangaroos can be of benefit.

Figure 6.10 Cross-grazing Credit: Jane Myers

The advantages of cows are that their fencing requirements are similar to horses and, like horses, they can be easily controlled with electric fences. Due to the differences between the two species (horses graze with their front teeth, biting the grass off, cattle tear the grass off with their tongues), horses can graze close to the ground whereas cattle need longer grass. This means that cattle tend to eat what horses leave. On a small property it usually works best to either put a cow or two in the paddock after the horses to clean up or to graze the cows with the horse herd moving around with them. If and when you have too much grass and your horses are in danger of getting too fat, cows can graze the paddock in front of the horses to reduce the available feed. The disadvantages of using cows for cross-grazing is that cows eat a lot of feed, this can be offset to some extent by using one of the very small breeds such as Dexter's. Cows' hooves also damage the land in much the same way as horses and they tend to be hard on the fences. Only use cattle without horns.

Sheep and goats will tackle most weeds found in horse paddocks, in particular Paterson's Curse, which is a big problem in southern Australia. Also sheep and goats are less susceptible to the poisonous effects of weeds such as fireweed. Goats are particularly good at eating gorse and blackberry and have the advantage over sheep in that most breeds do not require shearing, however, some breeds of sheep do not require shearing either (Wiltshire Horn). Larger goats are easier to keep in the paddock than smaller varieties of goats and sheep. As well as requiring shearing (most breeds of sheep), electric fencing does not always control these animals and they may need netting to keep them in. Beware of foot rot in prolonged wet weather.

Lamas and alpacas make good animals for cross-grazing due to causing less damage to the land with their feet than hoofed animals (they have pads rather than hoofs). They are placid and make good pets and even though they require shearing you may be able to get a local breeder to do it for free in return for the fleece. You may even want to do this yourself and keep the fleece for your own use. These animals also have the unusual benefit of being able to protect goats and sheep from predators and are being used for this purpose on farms.

In Australia you may already have small numbers of kangaroos visiting your property from nearby areas of bush. Kangaroos are gentle on the land and tend to eat the grass that horses leave. Like sheep, their manure is in small packages that are deposited evenly around the land. In order to preserve your fences provide a gap lower down in the fence large enough for them to get in and out. This will mean that you cannot graze sheep or goats in these paddocks, as they will use this gap also.

Emus and ostriches can be used as cross-grazing animals. The benefits include: their feet cause less damage than those of hoofed animals; they produce good manure; they scatter horse manure while searching for seeds; eat some harmful pasture insects and weeds; and they habituate horses to large scary birds. Emus and ostriches can usually be kept in a paddock with horse fencing. You will need to check your local council for regulations on keeping these birds.

Fowl are also good for pasture although there is a limit to how much grass and weeds they can eat. Fowl are good in conjunction with other cross-grazing animals as they scatter manure (searching for seeds) and eat weeds. Geese and ducks are good for trimming lawns. Geese have the added advantage of acting as an alarm service. They will let you know when you have visitors. Fowl must be supplied with their own dam or pond and prevented from using the dams that supply the other animals' drinking water, as their droppings will cause algae.

Borrowing or sharing animals with neighbours for cross-grazing can sometime be a good option so that you only have them on the property for a part of each year.

If you do decide to purchase other animals for cross-grazing, you will need to purchase a good book on the care of these animals also.

Strip grazing

Strip grazing is a system of grazing that involves using an electric fence to monitor how much the animals eat each day. This system can be used in conjunction with rotational grazing, for example the animals are still rotated around paddocks but are strip grazed across each paddock in turn. This method results in more even grazing as the animals move slowly, day by day, across the paddock rather than eating what they want and trampling the rest. It is more labour intensive than just turning the animals into the whole paddock because the fence must be moved on a regular basis so that the grazed area does not get too short. This method is especially advantageous for use with horses that put weight on too easily and are at risk of associated conditions. With this method the horses get a fresh but controlled amount of feed each day.

Other advantages are that horses are less likely to run around the paddock due to the smaller available paddock size and that if you are picking up manure it is easier to both find and pick up. This method requires portable electric fencing. This can work with either a portable energiser or, if the perimeter fence is electric, the strip fence can be joined into that.

Limitations to this system can depend on where the paddock water is positioned. The water may need to be transported to the section that the horses are using.

Strip grazing allows for a high level of paddock management and should be considered when the paddocks have reached a level of high yield.

Block grazing

If the paddock is going to take several weeks to graze using strip grazing then a second electric fence can be added behind the horses so that they cannot go back over the area that they have already grazed. This area can then be maintained by harrowing and mowing.

Figure 6.11 Strip grazing Credit: Jane Myers

7

Horse facilities

Horse facilities can range from basic to elaborate. There are certain facilities that are essential and others that are less so, depending on your needs and wants. Even though many different facilities are described here you do not necessarily need all of them. The most sensible thing to do is to start with the essential facilities and then add the others as and when money becomes available.

Aim to build strong and safe facilities that will last a long time. Economising on quality is false economy. Horses are very hard on facilities and you will have to rebuild them later on if you build inferior quality ones in the beginning.

On small properties space is at a premium so where possible facilities should be able to be used for more than one function. An example of this is an arena can be used for its intended purpose but if fenced can also be used to turn horses out into for exercise when pastures are indisposed.

So what should you plan to build first? For the sake of the welfare of your horse – water, food, companionship and then shade/shelter are essential in that order. For the sake of your property – protection from land degradation and enhancement of the environment are priorities. For your own sake – somewhere safe to work and handle your horses may be a priority. Taking these factors into consideration, strong safe yards that either have their own shelter or are attached to stables should be at the top of the list. These yards can be called sacrifice yards because the space that they take up is sacrificed so that the rest of the property can benefit.

Sacrifice yards

Probably the most important facility you will build in terms of efficient management of your land will be the sacrifice yards. This will mean that you will be able to control the amount of time that your horses spend on the land thus enabling you to manage the property in such a way that further down the track grazing can be maximised. This may

Figure 7.1 Sacrifice yards off stables Credit: Jane Myers

sound rather confusing until you understand what will happen to your land if you allow horses to have free range access to it on an ad hoc basis (see p. 90).

Other purposes that these yards serve are that they confine manure deposits and weed seeds, they provide a place to feed and water horses and they can also be used to confine horses that get too fat.

Each horse should have an individual yard unless they are very quiet and well behaved when together (for example a pair of very old horses). In the confines of a yard even the best of friends can injure each other, especially at feed times.

For the sake of good property management, yards should be built irrespective of whether stables are built or not. They can either be located separate to a stable building or directly on the outside of stables or indeed instead of stables. If the yards are not attached to stables then they will need shelter (see p. 126). If you are planning to have stables then it makes sense to have yards attached (rather than in a separate area), as this will save a lot of time that otherwise would be spent moving horses. It will also save doubling up on buildings, i.e. building stables and shelters. Also stables with yards attached are far better for horse health and wellbeing than those without (see p. 20).

Surface

The yards should be located on fairly level high ground (2 m above the highest water level in the area) and usually at least 50–100 m away from a waterway (check with your local council). The subsurface can be either non-permeable or permeable. Ideally it should be non-permeable to prevent leaching of nutrients into the underground water table. Compacted limestone will achieve this. In sensitive areas (i.e. near a watercourse) it may be necessary to lay a concrete base or use a rubber matting/concrete or limestone combination to prevent any leaching. If rubber mats are used you will need to place absorbent bedding in one area to soak up urine. The horse will tend to use this area because horses do not like to urinate on a hard surface as it then splashes on their legs. Other possibilities for a subsurface are to compact the existing soil or excavate this soil and replace it with layers of progressively finer material starting with rocks and ending up with gravel.

The subsurface will require a top surface that can cope with wet weather. Top surface materials can be various types of sand, sawdust, shavings, pinebark, shell grit, or fine gravel. What you use will depend on factors such as climate, and local availability of materials.

The yards should have a slight slope (2–4 per cent) for surface run-off. A buffer zone of grass and bushes can be created around the yard area that will act as a filter for any run-off and will help to retain the surface. Sleepers or logs can also be used to hold the surface. Run-off water from other areas should not be allowed to run across the area so water tanks should be fitted to any shelters or stables. Also drainage channels should be installed around the area to channel water around rather than through the yards.

Size and shape

The yards can be either square or rectangular (or any other shape for that matter). Long narrow runs are not ideal but are sometimes unavoidable when they are attached to stables. Aim to make the yards at least 7 m wide. An option for yards attached to stables is to have one larger yard per two stables that the horses can use for part of each day/night on a taking turns basis, or if space and budget allow have stables that are double the standard width. Another option, if you are planning to build four stables, is to have them in a U-shape giving more room around the outside for yards (see p. 131).

A common mistake is to make the yards too large or too small. If they are very large they are expensive to maintain, i.e. more surface material and more fencing and they use up too much space that could otherwise be dedicated to pasture. If they are too small or narrow the horse is too confined and cannot move around much or roll safely, and if they are next to another yarded horse one horse can bully the other. Aim for between 50 and 100 m^2.

The smaller the yard, the more exercise the horse will need to make up for the lack of space. Horses are not designed to stand still for long periods of time and suffer health problems if forced to do so.

Fencing

All fences on a horse property should be strong and safe. However, the more confined a horse is, the more it comes into contact with the fence. The strong and safe rule is especially true for fencing yards. Plain board fencing (post and rail) is the most common form of fencing for yards, however, they tend to splinter if horses run into them and this can sometimes injure horses. Wooden fencing also tends to be high maintenance and can soon start to look unkempt unless regularly serviced. Electric fencing can be incorporated and is ideal in all but small narrow areas which can result in the horse not being able to move freely without fear of touching the fence. Steel yards are a very suitable option and tend to be safer when horses are on either side of a fence (such as with yards attached to stables). With these an electric fence is not needed and horses can interact reasonably safely with one another over the fence. Mesh fencing should only be used if it is strong enough for horses and the gaps are

Figure 7.2 All steel sacrifice yards with shelters Credit: Annie Minton

small enough to prevent a hoof from getting caught. The fence should be 1.4 m or higher (see p. 151).

Gates should swing both ways, should lie flat against fences when open and be free of any projections. They should also be wide enough to allow machinery into paddocks or yards for topping up the surface and maintenance work. When closed, there should be no gaps between the fence and the gate that the horse can trap its head in (see p. 170).

Training yards

Training yards can be various shapes and sizes. Generally the main difference between a training yard and an arena is size (arenas are larger). A training yard is fenced with a high fence (higher than 1.8 m), whereas an arena can be left unfenced or usually has a lower fence.

Figure 7.3 A portable steel round yard Credit: Jane Myers

Figure 7.4 A combined round and square yard made from light weight portable steel panels Credit: Jane Myers

Figure 7.5 A rubber-lined post and rail round yard Credit: Jane Myers

Size and shape

Round yards are often used for starting/breaking horses in. For this reason they may be small, i.e. about 11.5 m diameter. As this is only one small period in the training of a horse, a yard this size is too small for general use. So, unless you are a professional trainer who uses it for this purpose regularly, a more useful size for a round yard is between 18–22 m in diameter.

Some people find a square yard of similar dimensions much more useful because then there are straight edges for certain exercises, and the yard can easily be made round if necessary by using panels or rails to temporarily round the corners off.

Yet another alternative is to have a yard that is a combination of round and square with one or two right angle corners and the rest rounded off (see Figure 7.4).

Surface

The criteria for the subsurface of a training yard are the same as for an arena (see p. 123). The surface considerations are very similar, although grass is not usually an option as it will wear off much more quickly in a yard than an arena. A good training yard surface drains well, gives good traction and has a low dust factor.

Fencing

Training yards can be fenced in hardwood or steel. One option is a portable system using purpose-made steel panels that are fastened together with pins. Portable panels are very useful as the shape and size of the yard can easily be altered to suit your needs.

Round yards can be fenced so that the walls/fence lean out so that the rider is less likely to bang their legs. The smaller the round yard, the more this becomes important. However, this style is more difficult to build and maintain. Round yards are more commonly being built with vertical sides for ease of maintenance and reduced initial cost.

The height of the walls depends on the type of horses being worked. To be safe, the top, or baulk rail, should be at about waist level when the rider is mounted (about 1.8–2 m). Make sure the posts are the same height as the top rail as projections can be dangerous.

Figure 7.6 A rubber-lined covered round yard Credit: Annie Minton

Figure 7.7 A rubber-lined steel training yard Credit: Annie Minton

Any rails must be on the inside to prevent riders (and horses) from banging into them. The posts can be set 1.8–2 m apart, but closer or further does not really matter unless it affects the fence strength. If you are using railway ties (sleepers) at the base of your yard you will want to space your posts accordingly. Sleepers can vary in length so check before installing your posts.

The walls can be solid (rails or rubber), which some trainers prefer as it prevents the horse from being distracted. However, solid walled training yards can be dangerous if the handler needs to get out quickly. They cost more to build and also concentrate the heat, reduce airflow and prevent the sun from getting in to dry the surface after wet weather.

Open style post and rail or pipe/panel yards are cheaper to build and to maintain. The bottom rail should be no less than 30 cm from the ground to reduce the risk of leg injuries to horses. Even better is to fit a rubber strip around the bottom edge.

Figure 7.8 A post and rail round yard Credit: Gary Blake

Figure 7.9 Building an indoor round yard Credit: The Horse Shed Shop

Figure 7.10 An all-weather covered round yard Credit: Annie Minton

A novel, inexpensive and ecologically friendly round yard can be built with old car tyres. These are stacked around a stay pole (one per stack) and are filled with sand or soil. Make sure you use tyres that are all the same size for a neat finish. Tyres act as a good buffer, are not see-through and will last indefinitely. Creeping plants can be grown in the tops of the tyres for a better look. Getting tyres is not difficult as tyre replacement companies are usually more than happy to give them to you. Just remember that if ever you want to get rid of them you may need to pay for the rubbish tip to accept them.

An added luxury is to have a roof on your training yard. This enables you to work in all weathers but of course it is quite expensive. A roof is something that can either be built in right from the start or be added later to an existing yard.

The gate should be able to be opened when you are mounted, so that you can ride in or out of the yard. Therefore the latch should be operable from both sides. The gate should be flush with the walls of the yard, have no gaps between the post and the gate when shut, have no projections and swing both ways. It needs to be wide enough to get machinery into the yard for topping up the surface and maintenance work (not less than 3 m). Alternatively two gates can be installed, the wider one for machinery and a narrower one for entering the yard with a horse (see p. 170).

Arenas

Riding arenas are classed as a luxury feature for some people and absolutely essential for others. It all depends on what you do with your horses, how much space you have available and whether you can afford one. An arena can be used for more than one function which may be a deciding factor when planning to install one. For example if it is fenced it can also be used as an occasional turnout area when the paddocks need to be rested.

Figure 7.11 An arena being used for exercise Credit: Jane Myers

Size and shape

A standard dressage arena is 20 m × 40 m, an Olympic sized dressage arena is 20 m × 60 m, an arena for jumping or western pursuits (such as roping) will usually need to be at least 40 m × 60 m and may need to be as large as 60 m × 100 m. It all depends on how seriously you follow your chosen pursuit (i.e. higher jumpers need a larger arena). On a small property this is a lot of space, so think carefully about how much you will use it.

Unless you are already very experienced in your particular pursuit, it is important to speak with professionals about what makes a good arena. In particular the size, surface and fencing requirements change across different disciplines. Also check with your local council prior to construction as there are often considerable earthworks involved that could impact on water catchment, protected trees or have other environmental implications. Neighbours should also be consulted. Council regulations may limit the arena to 'personal use only', which may prohibit use of the arena for commercial purposes such as coaching visiting riders.

Surface

There are many arena surfaces available that range from various types of sand, hardwood sawdust or shavings, rubber composite surfaces and commercially produced surfaces that are usually a mix of various materials.

If using sand, use a firmer type rather than deep soft sand. A deep sand surface is not good if you are planning to work horses for reasonable periods of time. This kind of surface puts too much strain on tendons, as it is very difficult to move in. On the other hand, a too hard surface can cause concussion injuries to legs.

The selection of surface will depend on many factors; some are listed below.

- Your budget.
- The availability of materials in your area.
- The climate, i.e. how much rain, sunlight/heat and wind the surface will have to contend with. High prevailing winds will bias surface selection towards the

Figure 7.12 A 60 m × 20 m dressage arena Credit: Equestrian Land Developments

heavier materials, while high levels of sunlight and heat will bias selection away from rubber composite surfaces that are more prone to UV degradation.

- The availability of water. Dust can be a major problem with arena surfaces and can alienate otherwise compliant neighbours. Unless you have lots of water available, choose a surface that produces as little dust as possible.
- How much you will be using the arena will also help to determine the best surface. For example you may get away with a cheaper option if you will only be using the arena occasionally.
- Your chosen pursuit: different disciplines have different requirements for surface so if you are not experienced speak with professionals first. For jumping, some cushioning is required but the surface should not be deep. Western events, such as cutting and reining, require a firm base with a thinner top surface that will allow slide. Dressage requires cushioning without too much depth. Rarely is the existing earth any good unless it happens to be predominantly sand. Soil will be too hard and dusty in dry weather and slippery/muddy in wet weather.

A good alternative is to have a flat grassed area that can be used as an occasional arena and also as another paddock. This area can be grassed with particularly hardwearing grasses such as kikuyu. It is still better to have the area professionally excavated so that it has the correct slope for drainage and the surface is smooth. The initial cost of this type of arena will still be considerable depending on how much earthwork needs to be done. However, once established, the arena will be aesthetically pleasing and add to rather than detract from your available grazing. Well managed grassed arenas also have a low dust factor that will keep your neighbours happy.

Costs

Installing an arena is an expensive exercise. The total cost will obviously vary depending on amount of earthworks, surface preparation, type required and surface. For example a 60 m × 20 m arena may cost upwards of $25 000 if professionally installed. A larger arena will obviously cost more. It is usually better to use a company that specialises in arenas because it requires specialist knowledge to construct one. As with many things, you get what you pay for and companies that specialise in arenas should provide a guarantee on both the surface and the base.

Construction

Some arenas are constructed so that water travels through the surface into drainage underneath. Others are designed so that water runs off the arena, in which case the arena either tilts slightly from one end to the other, or from one diagonal corner to the other or has its highest point down the middle and falls to either side. When done correctly this slope is not discernable to the eye. The company constructing your arena will have their own methods and will be aware of local factors. Arenas that are designed to have water running off the top should have a buffer of vegetation around the outside to hold the surface in and to filter run-off water.

Figure 7.13 An arena under construction showing new turf around outside edge
Credit: Equestrian Land Developments

Figure 7.14 PVC rails and wood posts in an arena fence Credit: Jane Myers

Fencing

An arena can be either unfenced or fenced. If fenced it needs to be safe for both riders and horses, secure, highly visible and preferably low maintenance. If you plan to also use it for occasional turnout for horses, the fence must be high enough and strong enough to keep them in. Because an arena is a small area and therefore horses are more likely to challenge the fence, consider a minimum of 1.4 m for a 15 hh horse. An arena that is used for teaching beginner riders and children should be securely fenced. If the arena is to be used for riding only and the riding does not include jumping then a fence of 1.2 m to 1.4 m is fine. Arenas that are used for jumping should be either left unfenced or fenced with a minimum height of 1.4 m. The fence should either be higher than the height of the jumps or the jumps should be placed well away from the fence. A professional show jumper would usually have a very large arena or paddock for jumping large jumps.

Fence material options for arenas include post and rail, steel pipe/panels, mesh and so on, the main considerations being safety for riders and horses. This eliminates electric fencing, steel pickets or any other fencing material that could cause impalement (see Chapter 8). The considerations for gateways are the same as for training yards.

Shelters

Yarded horses need shelter from the hot sun and from strong winds and rain. Most horses will use a shelter more for the purpose of shade and to escape flies in hot weather than they will use them in cold weather. In fact cold still weather does not bother horses in the least if they are healthy; it is driving wind and rain that they strive to escape from.

If a shelter is to be provided, open-sided ones are more acceptable to horses, preventing them from feeling claustrophobic. Millions of years of evolution have taught horses that to have no retreat is to be in danger. For this reason some shelters that look perfectly acceptable to humans, because they are enclosed, warm and dry, are virtually ignored by the horse who stands outside the shelter at the front using it as a windbreak only.

Figure 7.15 Double shelter across two yards Credit: The Horse Shed Shop

Figure 7.16 A shelter/stable Credit: Jane Myers

Horses actually prefer natural shelter such as trees and bushes. However, if these are not already present they will take time to grow and temporary shelter will need to be provided. This can be in the form of a simple roof such as a secure shade sail or tarpaulin. Trees need to be protected both from being chewed and from compaction at the base, so they need to be on the outside of any yards in order to survive. Deciduous trees that have leaves in summer but lose them in winter work well by giving shade in summer and allowing winter sun into a wet area. These trees are slower growing than natives and will not give shelter against bad weather in winter. They are best used as an addition to other shelter arrangements rather than as the only form of shelter.

Man-made shelters should be large enough for the horse, be easy to get in and out of, have no projections, be waterproof, cool and dark and be situated in a high and dry spot. Make sure the shelter is placed to take account of hot summer winds, wet winter winds etc. A shelter in the Southern Hemisphere should face east if possible, and if not, then north. It should also be either right up to or well away from the fence and it must be strong, as it will get rubbed by the horse.

A man-made shelter can consist of anything from just a simple roof made from a shade sail/tarpaulin/corrugated iron etc. to a building with between one and three walls. In hot climates steel shelters that have sides on them can be too hot. However, in colder windy climates sides are usually essential. A 'bus stop' style shelter which consists of just a roof and a back wall gives protection against the elements without being too enclosed. Another form of shelter is to put a large roof over the corners of four yards, this way horses can stand together while in the shade. Companionship is so important to horses that they will forgo comfort, and even food, in order to be near each. This kind of shelter can have either open sides or filled-in sides depending on the climate (see Figure 7.17).

The range of building materials for shelters is the same as for stables, and like stables any walls should be lined. Make sure that roof overhangs on shelters do not have sharp edges that can injure a horse. The internal height of a shelter should be 2.75 m or higher.

Shelters will need a surface and this can either be the same as that used for the enclosing yard or you may want to surface the shelter with rubber mats and use the area

Figure 7.17 An open-sided shelter for four Credit: Jane Myers

for feeding. The mats then reduce the amount of sand taken in by the horse if the yard is surfaced as such.

Shelters in paddocks

If you are striving to manage your horses and your property sustainably, then building shelters in paddocks is not usually necessary and can actually lead to increased land degradation as horses track backwards and forwards to them creating compacted paths/mud etc. Unless you have abundant grazing available (this is unlikely on a small property all year round), the horses should either be grazing or in sacrifice yards. You must aim to minimise the time that horses spend loafing in the paddock as this is one of the main causes of land degradation. If you do build paddock shelters or they are already situated in your paddocks, then enclose them in a yard. This then becomes a sacrifice yard and can be used as such.

Aim to increase natural shelter in paddocks by planting trees and bushes around the outside of paddocks because natural shelter has other benefits beside simply shelter/shade (see p. 175).

Stables

Stables are enclosures for horses that have four walls and a roof. It is very important that stables are well designed for optimum horse health and ease of use. Stables are actually built more for human comfort because horses are healthier if they are outside as much as possible. If a horse cannot live at pasture full-time then the next best thing for both mental health and physical health is a large yard with good natural or man-made shelter. Stables allow you to handle your horse in comfort by protecting you from the elements while you are doing so. Even well-designed stables confine horses more than they should and challenge a respiratory system that is not designed for indoor living. Keeping these facts in mind will help you to design stables that work well.

Figure 7.18 Four stables in a square, with grassed, steel fenced large yards
Credit: Annie Minton

Figure 7.19 back-to-back rows of stables with narrow passages up the middle. Horses enter and exit through the yards
Credit: Annie Minton

If you decide to have stables, there are many options open to you. You can have them built by professionals therefore saving you time and labour but usually being the more expensive option. The other option is to build them yourself, which will obviously be a labour-intensive procedure but can be rewarding and cost effective if you have the skills. Either way you need to have a good understanding of what constitutes good stables.

If using a builder, find someone that has been recommended by someone else or use one of the commercial stable building companies and remember that the cheapest quote is not necessarily the best. You need to check how many fittings they plan to use, the quality of the fixtures and fittings (which must be heavy duty with horses otherwise they will need to be replaced later), and how strong the building will be. Builders that do not have experience of building stables tend to underestimate the strength of horses and the amount of wear and tear they give.

If building the stables yourself, a pole building may be a good choice as this then negates you having to make or purchase a frame for the building. Alternatively you

Figure 7.20 A pole building under construction Credit: Jane Myers

could have a frame (wood or steel) erected by professionals and then finish the rest of the building off yourself. Yet another option is to have a shed erected (or buy a kit shed and erect it yourself), and then build stables inside the shed. It is possible to buy prefabricated sections that will turn your shed into stables or you can make the interior walls and partitions (see Figure 7.21).

Stables should be built on high ground. Ideally there should be a 2–6 per cent slope away from buildings for surface drainage. Buildings should be on a base 20–30 cm above the outside ground level. A diversion ditch should be dug around the back of the building if it is on a hill. Have the subsoil evaluated before building. Sand and gravel are better than clay for buildings. If the subsoil is mainly clay it may be necessary to have the area excavated and filled with rocks and road base or limestone. This will need to settle for several months or even longer before building.

Termites are a potential problem in certain areas. Termites will eat many materials so if they are a risk in your area choose termite-resistant materials such as concrete and steel. If termites are present in your area be vigilant and check for termite activity regularly and have your buildings checked by professional pest controllers periodically if you are not sure.

Figure 7.21 A professionally erected L-shaped shed and home-made stable (interior) Credit: Jane Myers

Figure 7.22 One stable and one combined feed/tack area with post and rail yard lined with mesh Credit: Jane Myers

Figure 7.23 A converted dairy that now includes four stables, a feed room and a tack room Credit: Jane Myers

Permits for buildings

Permits will be required for any significant sized buildings on your property. As requirements vary, you should check with your local council for their current conditions and requirements for permits. In order to get a permit you would usually need to fill in an application form and submit a site plan along with an engineer's computations, builder's specifications (structural details) and the builder's plans showing the floor plan, elevations, footings, etc. If you are using a professional to erect the building for you then they may do this as part of the service, but it is worth confirming this.

Stable design

Check out as many designs as possible before building or deciding what to have built. The design and arrangement of facilities in the stable can have a big influence on stable management. Facilities that are well planned and well designed make horse care much easier and more enjoyable. Allow space for movement of horses, wheelbarrows and maybe small machinery such as a front-end loader.

As well as individual boxes, a stable complex can include other areas such as a wash/vet area, tack storage, kitchenette, laundry, feed storage, separate tie-up areas, workshop, toilet/s etc. The first things to consider are: *How many stables are needed?* and *What will they be used for?* Then decide what other areas you are going to have, if any.

Some of the different styles of stables are described below.

- A straight single row of stables with either no overhang to the roof, a roof overhang (for some protection from the elements) or a full veranda that can be used for various purposes as well as protection from the elements.
- A double (back to back) row of stables with or without the same overhang/veranda options. As with the single row the horses have access to fresh air.
- Other variations are L-shaped or U-shaped rows which increase ease of use (for humans) without compromising on ventilation to the extent that an enclosed building (barn-style) does.
- Barn-style stables where two rows of stables face each other with a central aisle (breezeway) down the middle.

Barn-style stables are very popular because they allow stable chores to be done out of the weather. They are more costly than stables that are in a row and from the horse's point of view barn-style stables are the least desirable as they have the poorest air quality and the view is only of the inside of the barn itself. These factors can be greatly

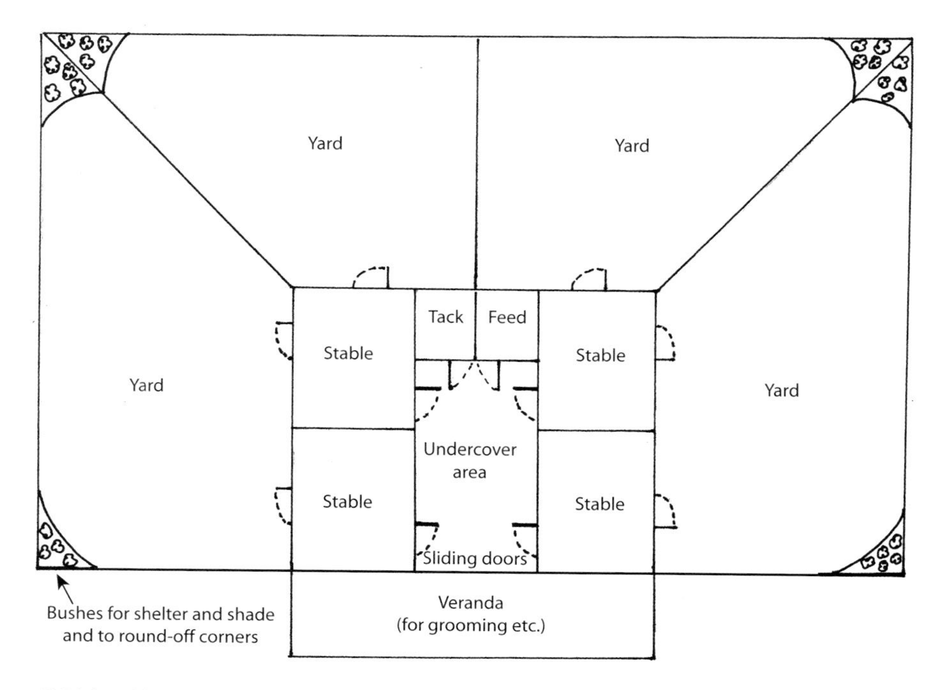

Figure 7.24 A stable design showing four stables with sacrifice yards attached, tack and feed room and undercover areas for grooming etc.

Credit: J & S Doig

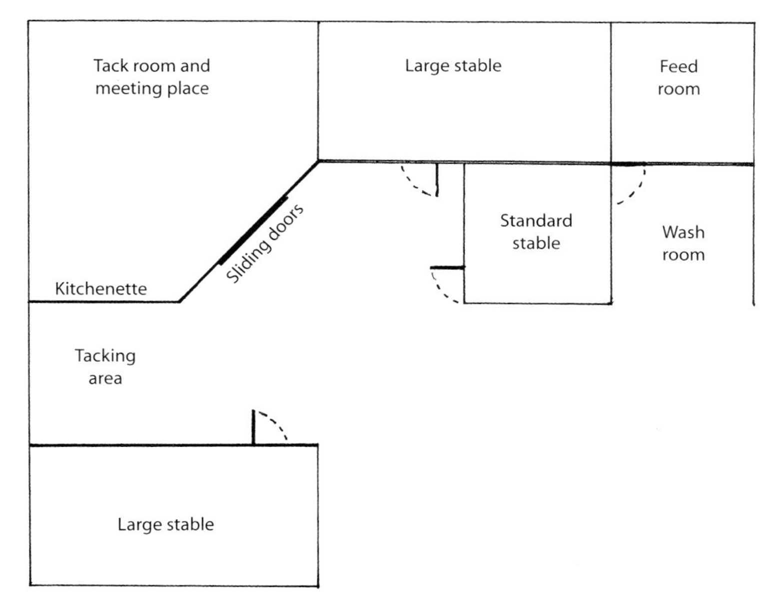

Figure 7.25 A stable design showing two large and one standard stable, with large tack room or meeting place, and tacking area. A lean-to on the side of the building can be used for feed storage and horse washing. This design could also have yards leading off each of the large stables. If could also incorporate more stables by reducing the tack room and/or dividing the large stables. Credit: Jane Myers

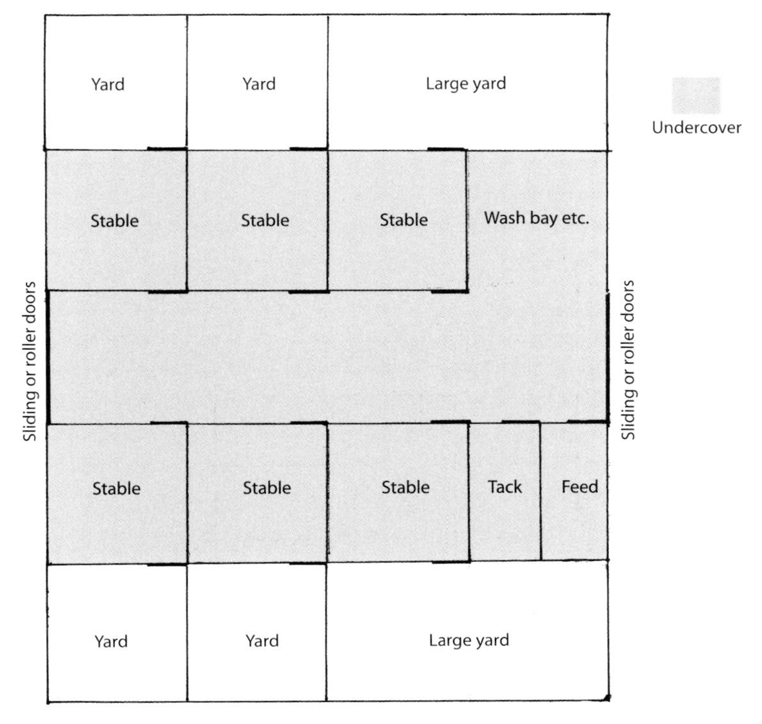

Figure 7.26 A barn-style design showing six stables, a tack, feed and washing area, and four small and two large yards outside. All doors are sliding. Credit: Jane Myers

Figure 7.27 Barn-style stables with yards nearby

Credit: The Horse Shed Shop

Figure 7.28 Window with shutters

Credit: Annie Minton

improved upon by having permanently open windows (or open stable top doors) to the outside of each box. There should also be more than one exit for fire safety, and these exits should be as far apart as possible (i.e. each end of the breezeway). Additional exits should be added down the sides of the building if there are more than 10 stables in a row. The breezeway needs to be wide enough for machinery (for moving feed/mucking out) and for moving horses. The flooring should be non-slip. A good width is about 3.6 m.

Roofs

The choice of roof materials includes any material that can be used for a domestic building. Some of the options are tiles, corrugated steel and shingles. Factors to consider are aesthetics (including in relation to the other buildings), cost, strength, safety (including fire safety) and ease of maintenance.

The roof should have a minimum inside height of 2.75 m for ventilation purposes and head clearance. Higher roofs give even better ventilation and are cooler in the heat. A common roof material is steel which can be either silver coloured (galvanised or zincalume®) or one of many other colours (i.e. colorbond® /colorsteel®). Many coun-

Figure 7.29 Clearlite panels over the breezeway

Credit: Jane Myers

Figure 7.30 Barn stables with loft

Credit: Annie Minton

cils do not allow silver-coloured tin to be used for buildings because it can produce a glare, so check with them first.

Although it is not essential, roofs can be insulated so that temperature extremes and condensation are reduced. This will also help to deaden noise in hailstorms and heavy rain. This insulation material will need to be held up with mesh or similar material, otherwise it will sag over time.

Clearlite panels (transparent panels) provide extra light into buildings that are enclosed and save on using as much power for lighting; however they also let in heat. In hotter climates use them only in the breezeway (rather than directly over stables) and limit their use to approximately one every 7 m of the length of the breezeway. Also position them on the side of the building that gets more shade than sun to further reduce heat.

With barn-style stables the roof can be a low or high loft style. High loft styles can have an upstairs floor which can be used for extra storage or even accommodation (check with the local council).

Roofs should be fitted with gutters and spouts and the water diverted (either by surface or subsurface drainage) so that it is taken away from the building without causing wet areas and does not run across yards. Water should be collected in water tanks whenever possible. The overflows from tanks will also need to have correct drainage to take surplus water away.

Ventilation

The pitch of the roof is important with regard to ventilation. A higher pitched roof that has ventilation outlets positioned high up will draw air upwards and outwards.

Open row stables have better ventilation than closed buildings such as barn-style stables. Stables need at least 1 m of free air flow above the head level of each horse; therefore 2.75 m is the recommended roof height and higher is even better. Stables that are part of a complex (barn-style stables) can have poor ventilation in the middle stables compared to those on the ends. Open windows/top doors make a huge difference to air quality in this style of stables. In addition the roof will need to allow air to flow out so that the air is changing regularly. The emphasis should be on fresh air flow rather than draughts (see p. 20).

Figure 7.31 These barn-style stables have roof ventilation, side entrances and open windows in each stable, giving good ventilation Credit: Annie Minton

Walls and partitions

The exterior of the building may determine to some extent the interior. Some building materials will make up both the inside as well as the outside aspect, while some require lining. Whatever the material, the interior aspect should be strong, smooth and should not splinter if kicked. Some of the materials that can be used for building stables and lining walls (kicking boards) and partitions are shown below.

- Corrugated steel, which is quick to erect, and gives a neat and tidy finish. It can either be coloured or not coloured (silver). The colours can be matched to other buildings such as the house/garages. This is used as an exterior finish as it must be lined, otherwise a horse can put a hoof through it, which will result in injury.
- Flat galvanised steel sheets can be used for partitions between stables and as a liner for exterior walls made of wood or corrugated steel for example.
- Timber is very versatile as it can be used as an outer wall, in which case if it is thick enough (2 in or 5 cm) it will not require lining. It is also often used for lining outer walls (such as corrugated steel) and can be used for partitions. When used as a liner, planks should be 5 cm thick to a height of at least 1.4 m. The planks can either be fitted vertically or horizontally. They should be fitted flush to the floor so that a horse cannot get a hoof caught underneath, however, this means that the bottom plank (or the bottom edge of vertically fitted planks) will eventually rot unless they are protected in some way. Plywood sheets can also be used for lining which tend to rot unless they are treated, such as marine ply. Another problem with sheets is that if a horse does manage to kick through one, the whole sheet will need to be replaced. Wood is not recommended in areas that have a high risk of termites and exposed wood edges inside stables should be covered in steel to prevent chewing.
- Concrete is a very good material for building walls or lining stables, as well as wash bays, as it is easy to keep clean (and therefore reduces dust), is fire resistant, low maintenance, cool and strong. Concrete can be painted or left unpainted. Concrete can be poured into moulds to make up the walls (eliminating the need for an inner lining) or pre-made concrete wall panels can be used. Concrete blocks can also withstand rough treatment but will collect dust unless they are rendered.
- Commercially produced man-made weatherboards (i.e. Weathertex™ etc) are long lasting and more durable than timber. They are low maintenance but more expensive than timber. These can be used as an outer wall only as they will require lining to be strong enough for horses.

As well as the suggestions above, walls and partitions can be lined with rubber belting/matting to minimise injury to the legs of horses that kick out at the walls. However, this will add a lot of expense and should not be necessary with the average horse. Kicking walls is either due to frustration at being confined, in which case the horse is stressed and something should be done to enhance the environment of the horse, or the horse has learned to kick the walls or door at feed time and is being inadvertently rewarded for it by being fed on demand.

Stables should be either open above the height of the kicking boards or they can be meshed or barred. Stables that are solid from top to bottom, as well as being more

Figure 7.32 Stable showing high concrete walls with vertical bars Credit: Annie Minton

expensive to build, have reduced air flow and do not allow any interaction between occupants. Meshed or barred upper partitions allow some interaction and air flow between stables.

Bars should be no more than 7 cm apart to reduce the risk of a rearing horse putting a hoof through them. Alternatively leave the stables open above the kicking boards. This gives better air flow, is less area for dust to accumulate and allows horses to interact and groom each other over the walls. A good design that allows horses to interact but gives each horse protection when feeding is shown on page 20.

Stable size

The traditional minimum size for an individual stable (box) is 12 ft × 12 ft, which is 3.65 m × 3.65 m. A stable for a pony can be 2.4 m × 2.4 m. Large horses (more than 16.2 hh) and horses that have to spend a lot of time confined need larger stables, i.e. 4.3 m × 4.3 m or the stable should open out on to a yard. The stable can be as large as you like so, if the budget allows, build stables that are larger than the industry standards for the sake of your horse's comfort.

Smaller stables are fine if they are only for occasional use, such as feeding, and horses are only spending a few hours a day in them. In this case it is a good idea to make them with partitions that swing back against the wall or can be removed so that two small stables can become one large one if necessary, such as in the case of an injured horse requiring confinement.

Windows

Windows that open or are not covered greatly increase ventilation, and should be incorporated into individual boxes in barn-style stables for that reason. They must be barred if they have glass or Perspex in them or a better option is to have no glass or Perspex cover (with or without shutters), so that the horse can put its head out into the fresh air at any time. This also reduces boredom for the horse. They should be approximately 1.2m^2 and 1.3 to 1.5 m from the ground. If the stables have doors leading out to yards then either this door can be left open or just the top half of the door can be open if the horse is not being allowed access to the yard for some reason.

Doors

Individual stable doors should either open out or slide. Arguably they are safer if fitted to the inside, to prevent the possibility of a leg being pushed through any 'crack' between the door and wall of an outside-hung door. Better still is for the door to slide into a front and back casing. A door for horses that are 15 hh plus should be at least 1.2 m wide be 1.4 m high and have 2.4 m of head clearance.

Modern stables invariably have barred or meshed top doors that prevent the horse from putting its head over into the breezeway or aisle. The idea being that it prevents horses from performing the 'vice' called weaving. However, even if it prevents the horse from weaving (sometimes the horse still weaves anyway in the middle of the stable), it does nothing about the cause and actually increases stress levels in the horse by reducing movement and view. In stables where the general public are allowed access there is a case for preventing horses from putting their head over the door. In this case, this

Figure 7.33 Stable showing mesh above timber Credit: Jane Myers

Figure 7.34 Stable showing horizontal bars above wood Credit: Jane Myers

Figure 7.35 Barn-style stables showing open top doors Credit: Annie Minton

should be compensated for with an open window or open top door at the back of the box (see p. 138).

Roller doors can be fitted to the breezeway ends or twin sliding doors can be used. Large swing doors are generally not ideal as these are hard to handle if the wind catches them.

Doors to the feed rooms etc. can be swing, slide or roller doors. A useful option in a stable complex is to have an additional external roller door on a feed room for ease of access when feed is delivered.

Flooring

The flooring of a stable complex should be safe to use, have low dust properties, be relatively level and easy to keep clean. Floors in the breezeways and individual boxes can either slope slightly for drainage or be flat and rely on bedding/sweeping for getting rid of moisture. A slight slope in concreted boxes is useful so that most of the moisture in a stable is soaked up by bedding but the stable can be hosed from time to time when completely cleaned out. Underground drainage inside individual boxes is usually too problematic to operate. Because of the nature of bedding and horse waste, underground drains constantly block up. Any open drainage channels must lead to a septic or water purifying system rather than be allowed to run off.

A stable complex floor can either be concreted right through including the individual boxes, just in specific areas such as the breezeway and tack/feed/wash rooms or not at all. When concrete is laid it needs to be at least 100 mm thick.

Options for the floors of boxes (individual stables) include compacted earth, fine gravel (crusher dirt) or limestone, sand, or bricks laid on sand or concrete. The advantages of earth, gravel, limestone and sand are that they are cheaper, safer for horses to move around and lie on and they require less bedding than harder surfaces. The disadvantages are that they allow seepage of urine and manure (possibly into underground water), they can smell and they require periodic maintenance. Bricks on sand also allow seepage and can smell. They are very slippery when wet so deep bedding must be used.

Figure 7.36 Rubber mats Credit: Annie Minton

Concrete prevents seepage and is easy to clean but is expensive and can cause injuries if the bedding is not thick enough.

The most ecologically friendly and safest option for individual boxes is to lay rubber mats on top of compacted earth, gravel, limestone, bricks or concrete. Rubber flooring is generally made from recycled tyre rubber. Mats can be purchased that lock together or large mats can be sealed between the gaps and at the edges so there is no seepage into the ground and no smell. With rubber mats only a small amount of absorbent bedding is required to soak up liquids. Having rubber over hard floor surfaces such as concrete also reduces stress on the horse's legs.

Even though rubber matting/flooring has an initial high cost it has many positive features as listed below.

- A stable with a rubber floor requires less bedding, bedding is only required to soak up waste and to provide a bed. This means that there is less waste to get rid of and the stable can be cleaned out much more quickly.
- The softer flatter surface reduces shock to the horse's legs and provides an even surface for horses to stand on. This can reduce stress on horses' legs considerably.
- Rubber is far safer as horses do not slip as easily as on bricks or concrete. For this reason it is a good idea to have rubber on concrete walkways as well.
- A rubber mat prevents horses digging holes in the stable floor.
- The stable environment is less dusty due to less bedding. This advantage can be further enhanced by using the more expensive low dust varieties of bedding – as less is needed, the overall cost is reduced.
- The rubber mats are usually made from recycled tyres making it a good choice for environmentally aware people.

Rubber mats can be laid over a brick or concrete floor or a less expensive compacted earth, gravel or limestone floor. They can also be used for part of a yard surface (for example in the shelter) and they provide a good surface for feeding horses on especially in sandy areas.

The advantages of concrete in breezeways and other high traffic areas are that it is easy to keep clean, it can be swept or washed easily, it has low dust properties, its level finish makes it safe for people to walk on and it has a neat and tidy appearance. The

Figure 7.37 Concrete breezeway and rug racks ouside each stable Credit: Jane Myers

disadvantages are that it is expensive to install, it is cold underfoot and it can be slippery for horses (especially shod horses). When having concrete laid, make sure it is finished with a rough or patterned finish to make it safer. Other options for these areas include an earth, limestone or sand surface or rubber mats.

Stable fittings

Stables (and yards) can be fitted with automatic waterers/drinkers and swing-out feeders can be fitted to the front walls of stables. This means that horses can be fed safely and quickly from outside the stable.

There are pros and cons associated with automatic waterers/drinkers that should be considered if you are planning to fit them.

Cons

- The initial expense (they cost a lot more than buckets!)
- They require frequent checking for malfunction (which may mean that they stop working or don't stop working – and so cause floods)
- They must be cleaned regularly as they can collect manure and feed
- It is difficult to monitor a horse's water intake, which is essential if a horse is ill although it is possible to buy models that monitor consumption
- They can freeze in cold weather
- Some horses enjoy playing with them and can either continuously damage them or cause floods.

Pros

- They save on carrying heavy buckets of water
- They cannot be tipped over by horses (as buckets can)
- They give a constant water supply (as long as they do not malfunction).

Therefore automatic waterers can save labour and time but they are not completely maintenance free. The initial cost of installing them can be recouped by reduced labour

Figure 7.38 A pull-out feeder Credit: Annie Minton

Figure 7.39 An automatic waterer Credit: Jane Myers

costs in commercial stables but not so much in private ones. This puts them in the luxury category.

Rug racks can be fitted to the wall next to the door on the outside of the box so that rugs can be hung somewhere when not on the horse.

Tie rings can be attached to the inside of stables but these may not be necessary if you are going to use tying areas or the outside of the stable for grooming and tacking. Wherever tie rings are fitted they must be very securely fastened and the horse tied to string between the tie ring and the lead rope. This is because it is common for horses to pull a tie ring out of the wall when 'pulling back' and in doing so injure people or itself with the tie ring.

Hay racks inside stables keep hay off the floor. However, feeding hay on the floor allows the horse to eat with its head in the grazing position which drains the sinuses.

Figure 7.40 Tie ring
Credit: Jane Myers

Figure 7.41 Hay rack, feeder and waterer (filled by operating a lever outside the stable, therefore reducing the negative factors of automatic waterers)
Credit: Annie Minton

Wash areas

A horse wash area can either be indoors or outdoors. It is not imperative to have one (horses can be simply sponged off after work), but it can be a very useful addition. If you are building stables and have room in the building it can be situated there. On a small horse property it makes sense to have the horse wash area as a combined tying area/horse wash that can also be used by the farrier and for any vet work. A wash bay is usually the same size as a stable (3.6 m × 3.6 m). The flooring should either be rough textured rubber (smooth rubber gets very slippery when wet) or rough finished/patterned concrete.

The hose can be on a swinging arm so that washing both sides of the horse is easily done without having to drag the hose over the floor. A spray attachment on the end of the hose reduces water consumption and is easier to use.

In an indoor horse wash, shelves for storing washing products and tools can be fitted. They need to either be recessed or in a corner to reduce the chances of a horse banging them with its head. A sink is also very useful for washing horse gear.

Figure 7.42 A horse wash
Credit: Annie Minton

Hot water is another bonus. For this a domestic hot water system can be fitted and in indoor horse washes an infra-red heater is useful (but not necessary) especially in colder climates.

Water from a horse wash should drain into either a septic or grey water system. On a small property the domestic system may be able to cope with the water from the horse wash. This will depend on how many horses you plan to wash and how often.

Tying areas

Tying areas are required for grooming and tacking horses. Whenever possible horses should not be groomed inside a stable because of the dust factor (both for the one grooming and the horse). Grooming releases lots of hair and more importantly dander which will either immediately be breathed in by the person grooming or the horse, or will fall to the ground to be breathed in later when the horse lies down. Locate tying areas both outside and inside the building and use the inside ones only when necessary, such as in bad weather. These areas need to be reasonably near the tack room for convenience. A possible tying area is a veranda on the outside of a block of stables. This gives shelter and shade while at the same time reducing dust inside the stables. A veranda for

Figure 7.43 Tie-up areas Credit: Annie Minton

Figure 7.44 A feed room Credit: Annie Minton

Figure 7.45 A feed bin Credit: Annie Minton

this purpose needs to be approximately 3.6 m wide as any narrower does not give adequate protection from the elements.

Feed storage

It is convenient to store feed close to where it is needed. This will mean either having a feed room as part of a stable complex or in a building near the horse yards.

A feed room should have a smooth concrete floor so that it can easily be kept clean by sweeping or occasional hosing. It is vital that rodents do not get into feed due to the associated dangers of botulism and other diseases. The presence of vermin also encourages snakes, which can be a problem in Australia.

A feed room should be kept locked or very secure so that any escaped horses cannot get into it.

Feed can be stored in many kinds of container ranging from purpose-built feed bins to clean recycled drums made from plastic or steel. These are often sold inexpensively at feed stores. All feed bins must be vermin proof and waterproof.

The amount of storage space needed will depend on the number of horses to be fed, their feed requirements and how often the feed supply is to be replenished. Feed bins that are too small will need to be refilled every few days. Too large means the feed may become stale before you use it. Remember that each feed bin should be emptied completely before being replenished so that feed at the bottom of the bin does not become stale and unpalatable. Alternatively, when refilling, remove the old feed, put in the fresh feed and then place the old feed back on the top.

Hay storage should not be close to stables for several reasons. Large amounts of hay are a fire risk (both from self-combustion and external causes) and hay can create a dusty environment in enclosed areas. If possible, plan to build an additional shed that can be used for larger amounts of hay storage. It is better to bring only the amount need for each feed into a stable complex at a time. Or better still shake out hay in another area before bringing it into the stables to reduce dust.

Tack storage

A tack room can either be part of a stable complex or in a separate building, such as a shed, but it needs to be close to the tacking area. Storage for tack needs to be secure (tack is prone to theft) and dry. Leather in particular becomes mouldy very quickly if kept in a damp environment.

A tack room in a stable complex can either be a simple area for storing tack or a more elaborate area including a small kitchen/drinks facilities, heating, laundry, phone (office).

If space is a problem, a combined feed and tack room is an option with the main disadvantage being that the tack will tend to get covered in feed dust.

Hang tack on either purpose-built or improvised hangers and racks for ease of use and some protection against damp and rodents and insects. Tack and rugs that are not in current use can be stored in storage containers such as plastic ones that can be purchased cheaply from variety stores. This keeps it dry and protects it from vermin and insects until you need it.

Other buildings

A building such as a three-sided farm shed is very useful for bulk hay, float and farm machinery storage (such as a tractor and implements). If the building also has an enclosed room at one end (with or without a roller door), this can be used as workshop and for storage of tack and feed if you do not have a stable complex.

Lights and power

The stable, other buildings and outside areas should be well lit for safety and convenience. Each individual box should have its own lightbulb or strip. Light and light fittings should be fitted well out of reach of horses and there must be a safety cut-out switch fitted to the power box. Halogen, incandescent or florescent lights can be used in stables. Halogen lights last longer and are brighter than incandescent and tend to be more reliable than florescent.

Have lights fitted to the outside of buildings so that it is possible to walk to the buildings at night safely from the house.

Figure 7.46 A tack room Credit: Jane Myers

A more ecologically friendly alternative is to use solar power for the horse facilities. This is becoming cheaper to install and once set up is very cheap to run. Using solar gives you more flexibility when positioning facilities as it is not necessary to be near the main power source.

Any exposed power cables should be enclosed in conduit to protect them from chewing by rats. They should not be in reach of horses either.

You may also wish to have power points installed to your stables/other buildings so that you can clip horses/boil water etc. Power points can be fitted inside and outside buildings using heavy duty all-weather sockets. All electrical work should be done by a qualified electrician.

Plumbing

If you want to have running water to the horse wash, automatic drinkers and taps, you will need to contract a plumber to do this. At least one tap inside and one outside a stable complex is useful. Taps near the horse yards (sacrifice yards) will reduce the labour when carrying buckets of water.

8

Fences and fencing

Good fences are a must on a horse property. As well as keeping horses safe, they protect areas from horses and maintain boundaries between your property and your neighbours or your property and the road. Good fences reduce the chance of accidents occurring to people, horses and third parties. In addition, good fences are aesthetically pleasing and increase the value of a property. Having good fences does not have to be an expensive operation as the most expensive options are not necessarily the safest or the best.

There is no one kind of fence that is ideal for all situations, climates, type of horse or budgets. There have been many advances in fence types in recent years that have resulted in a wider variety of fencing which gives the horse owner plenty of choice, whatever the budget.

Many horse properties have several different kinds of fencing as different areas often have different requirements. The fencing used in a particular area depends on many factors including the paddock or yard size, the size and type of horses, whether you have other livestock as well as horses and your budget. Using different fence types in different parts of the property is often the best way of avoiding the disadvantages of certain fence types.

Safety, cost and appearance are other important considerations when designing fences for horses.

Fences should also keep intruders out. Dogs and children (whether they are your own or the neighbours') must be prevented from entering horse areas unless given permission (such as with your own children), because as well as not wishing anyone else's children or pets to get injured you may be legally liable if they do.

Fences are one of the biggest killers of horses but are only dangerous when the horse actually comes into contact with them. However, there are certain things that you can do to reduce the risks.

- Turn horses out together to reduce the incidence of horses walking fence lines and even challenging fences in an attempt to get to other horses. As well as being safer for horses, this reduces land degradation and allows better pasture management because rotation of paddocks can then be used as a management tool.

- Use electric fencing to keep horses away from fences as much as possible. Either using all electric fences or adding electric to existing and boundary fences keeps horses away from the fence and therefore away from danger.
- Train your horses to stay calm when they have things wrapped around their legs. This is a very important but often overlooked point. If a horse stands still rather than panics when caught up in a fence the damage can be minimal or even nil compared to the ripping injuries that horses sustain when they panic. If you do not know how to do this, contact a professional horse person who specialises in handling and particularly desensitising/habituating horses to accept scary situations.

Fence placement

The perimeter fence goes around the boundary of the property and of course there can be no flexibility there. However, before you erect any new boundary fencing, check the boundary (survey posts) so that you do not end up in a dispute with your neighbours.

The placement of the internal fences requires careful planning if the land is to be managed so that land degradation is minimised (see p. 65).

Fence visibility

Horses can, and sometimes do, run into fences. However, the greatest danger is when a horse is turned into a paddock for the first time or when they have been overconfined and fed high energy feed for a period before turning out. The colour of the fence is not important as a horse can see a dark fence just as easily as a white fence. The main thing is to make sure the fence presents a visible barrier to the horse. Obviously, a fence with widely spaced posts and only a few wires will be harder to see than a more solid fence. An existing fence can be made more visible by adding a white electric tape, PVC sighting wire or a rail to the top of the fence. Alternatively, a fence can be made more visible by attaching plastic bucket lids or plastic shopping bags to the top wire. Once a horse is aware of the fences these can usually be safely removed.

Fence dimensions

For average sized horses (14 to 16hh) fences should be a minimum of 1.2 m (4 ft) high but consider having higher fences for the perimeter fence i.e. 1.5 m (which is close to 5 ft) for any perimeter fence between the horses and a road. As well as being a better barrier, if they are taller it makes it harder for horses to lean over to graze on the other side of the fence which loosens and breaks fences very quickly. Whenever possible, prevent horses from doing this by either making the fence higher and impenetrable (i.e. mesh) and/or adding electric to the fence. Internal fences can be from 1.2 m upwards. Again there is no harm in having them higher than this if necessary.

Figure 8.1 A high steel fence Credit: Annie Minton

The spacing between rails, wires or pipes should be close enough (approximately 20 cm apart) to prevent a horse from putting its head through the fence or wide enough so that it can easily withdraw its head if it does (approximately 50 cm). With post and rail fences a wire can be put between alternate rails to prevent the horse from putting its head through. It is worth repeating – whenever possible, prevent a horse from putting its head through a fence or leaning over it as this is asking for trouble. If you can prevent a horse from even going near a solid fence by using an electric strand of wire then this is even better. The lowest element of a permanent fence should be at least 30 cm from the ground to reduce hoof and leg injuries. It can be higher especially on internal fences, i.e. 60 cm, depending on if and what other animals you are planning to keep as well as horses in which case mesh may need to be incorporated at the bottom of the fence. Electric fencing can then be used at horse height (near the top) to keep horses away from it.

Temporary electric fences are usually set lower (a tread-in post is approximately 75 cm) because horses do not tend to go near them and are therefore unlikely to step over them or lean over them as they will with an unelectrified fence. If a horse does get caught up in a single strand (of tape or braid) electric fence, it usually breaks which is far safer than a fixed fence.

Fences for yards should be a minimum of 1.4 m high (see p. 117). Fences for training yards should be approximately 1.8–2 m. The bottom rail should be no less than 30 cm from the ground to reduce the risk of leg injuries to horses. Even better is to fit a rubber strip around the bottom edge (see p. 120). Arenas can either be left unfenced or fenced with the height depending on what the arena is to be used for (see p. 126).

Post spacings depend on whether droppers/battens are used or not. With a post and rail fence the posts are usually placed approximately 2.4 m apart (about 8 ft) but for a wire fence the posts can be placed further apart and several droppers/battens used in between to hold the wires in place.

Generally speaking, a higher fence is safer than a lower fence and a fence that the horse cannot get its head through or lean over is safer than one that it can. Another generalisation is that the smaller the enclosure the higher and stronger the fence should be.

Costing fences

Another area of potential dispute with your neighbours is the subject of who pays what with a shared boundary fence. The cost of plain wire fences is traditionally shared equally between neighbours. However, if you want something more expensive, you will be expected to pay. Your neighbours may have no interest in upgrading the fences if they have no livestock, so in this case it may be better if you stand the cost yourself and keep them on side.

The cost of fencing varies greatly so it is best to shop around before buying. If you are erecting the fencing yourself, try to buy in bulk so that you can negotiate a better deal when buying materials. Likewise, if using a contractor, it is more cost effective to get it all done at once rather than bit by bit. When costing fencing you need to work out what the whole fence will cost per metre when erected so that you can compare different options. Don't forget to include all fixings and fittings as these make a big difference to the final price. You also need to take into account any ongoing maintenance further down the track. Fence types tend to fall into one of the following scenarios:

1. expensive to install but with very little ongoing maintenance
2. inexpensive to install with very little ongoing maintenance
3. expensive to install with high maintenance requirements
4. inexpensive to install with high maintenance requirements.

Types of fences

As there are many fence types available it can sometimes be better to combine more than one type within a fence, for example timber posts with a single top rail and wires below. This gives the advantages of strength and visibility, is more aesthetically pleasing than an all-wire fence but is not as expensive as an all-timber fence.

Keep in mind that internal fences and perimeter fences serve different purposes. The perimeter fence is in many ways the most important fence on the property as this is the last line of defence between your animals and the road or your neighbours' property. For this reason it should be the strongest fence on the property. It would not be recommended to use electric tape or braid only for this type of fence. However, these materials combined with more solid components such as galvanised wire, mesh, timber or man-made materials would be a very good fence indeed. On the other hand, the internal fences are there to keep horses in the correct areas but do not usually need to be as physically strong as a perimeter fence. A totally electric fence is fine in this situation, as long as it is properly constructed and maintained.

Timber fences

One of the more expensive fence types is a full timber fence. There are a few different types ranging from the very neat square-cut posts and flat rails (post and rail fence) to more traditional timber fences. These fences are made from rough-cut timber that is erected using no fixings as the rails are slotted into the posts. The latter kind of timber fence can actually be the more expensive of the two because a lot of labour is involved in making it.

Figure 8.2 A traditional post and rail fence Credit: Jane Myers

A post and rail fence consists of timber posts and up to five rails, depending on the height. When there are less than three rails the gaps may be made up with wires or mesh. Timber fences are expensive because of the timber rails and the higher number of posts that are needed as droppers/battens cannot be used with this type of fence. These fences are popular because they give the traditional look that is associated with horse properties. However, they tend to fall into scenario 3, being expensive to install with high maintenance requirements and they have a few other drawbacks as well, as shown below.

- They are susceptible to fire, rot and termite damage, although some hardwood timbers such as ironbark and treated pine are much better than others at resisting rot and termites.
- Horses like to chew them. This can be prevented to a large extent by protecting them with electric fencing, however, you need to factor this into the costs.
- Although often regarded as the safest fence, they are not as they do not give if horses run into them and they can splinter causing injuries. In addition, the fixings can cause injuries if they come free in an accident situation.

A post and rail timber fence can be painted white or stained black or brown if treated with creosote or old sump oil (usually free from garages). White painted fences

Figure 8.3 Timber post and rail fence Credit: Annie Minton

Figure 8.4 Examples of wire twitches on a combined timber and wire fence
Credit: Jane Myers

make a louder statement when new but quickly start to look unkempt unless regularly maintained. Treated fences blend better with the landscape and require less maintenance than white painted fences because they do not need attention as often.

For safety reasons, the boards of a post-and-rail fence should be placed on the inside of the posts to prevent horses from knocking the rails off if they hit them front on, or knocking their hips or other body parts on the posts if they travel at speed along the fence. Of course, this is not possible where only one fence separates paddocks. This risk is reduced when an electric strand of fencing is used in conjunction.

The rails can be 10–15 cm wide by 2–2.5 cm thick. Obviously, wider thicker rails are stronger and look better but are more expensive. They should be bolted (never nailed) with nuts and bolts or coach bolts onto the posts. All protruding fastenings should be countersunk to reduce the chance of injuries. Another method of fixing rails to posts is to use wire twitches.

Posts can be set 2.4 m apart at their centres. By buying the rails in 4.8 m lengths and alternating the joins between posts the fence will be stronger.

In short, a timber fence is often used as a front perimeter fence for its strength and looks with other types of fencing used elsewhere on the property. When timber fencing is used, consider using electric fencing in conjunction with it to increase its safety factor and its longevity.

Pipe and steel fences

Pipe and steel fences are expensive to erect but have many advantages including: not needing painting (unless you want to) so maintenance costs are low; they are versatile as they can be used as a paddock fence or for yards and arenas; horses cannot chew them; they are fire and rot resistant; insects cannot eat them; they are neat and tidy to look at and are very strong. The main disadvantage, besides the expense, is that they do not yield if a horse hits them.

Pipes come in various thicknesses with 40 mm to 200 mm being the most useful. The thicker gauges can be used as fence posts in conjunction with other types of fencing

Figure 8.5 Pipe fence  Credit: Jane Myers

Figure 8.6 A steel fence Credit: Annie Minton

materials such as wire. With a complete pipe fence the same thickness is usually used for the posts and the rails (see Figure 8.5). The pipes must either be welded to the posts or fitted with pipe clamps. Pipes can sometimes be obtained second-hand from scrap metal or second-hand building material yards. Rails should be set at the same distances as for rails on a timber fence. The costing of this type of fence all depends on whether you are able to buy the material second-hand or not. New pipe works out quite expensive.

It is possible to buy ready-made steel fencing either as separate components which are then welded together or in ready-made lengths (see Figure 8.6). This type of fence is expensive but very low maintenance and relatively safe.

Stone fences/walls

In some parts of Australia and New Zealand, stone fences/walls have been built for livestock. These fences were erected without the use of mortar, hence the original name for them (in Britain) is 'dry stone walls'. They were built a long time ago by pioneering immigrants in areas that had lots of stone for two reasons – to dispose of stone to clear grazing areas and to erect a fence at the same time. Stone fences are durable (they last indefinitely with some maintenance) and aesthetically pleasing. If they are already on your property, they should be maintained and cherished. However, to erect a new one today would be very expensive due to the large quantity of stone required as well as their being very labour intensive and requiring a high level of skill. A stone fence is not an option unless you live in a very stony area and/or want it as a feature, such as an entranceway to a property.

Mesh fences

Mesh is a popular choice of fencing for horses because it keeps all classes of animals in, intruders out and prevents horses from putting their head through the fence. Mesh can either be a dangerous or safe fencing material depending on how it is made and the size of the gaps in the mesh. Mesh is either spot welded, linked or woven with interlocking joints; whatever the type of mesh the holes need to be small enough to prevent a horse's leg going through (about 5 cm or less). Shod horses are more at risk with all types of mesh fences (and wire fences) because the mesh/wire can get caught between the hoof and the shoe (even if the foot does not go through). This can result in the shoe simply being pulled off or it can result in a serious injury to the hoof if the shoe pulls part of the hoof with it. This is quite a common injury to horses and therefore deserves serious consideration when choosing a fence type. Some types of mesh fence require a tight wire along the top and bottom of the mesh to keep the desired shape.

Two common types of mesh fences are 'ring lock' and 'dog fence'. Both have large gaps that a hoof can easily go though. These types of mesh are either spot welded or linked. Spot welded mesh can snap leaving sharp protrusions and does not work well on undulating country as it cannot flex and accommodate the dips and rises. Linked mesh, such as ring lock, is able to accommodate undulations. Both are unsuitable for horses if the horse is able to get close to the fence but are good for perimeter fences (and for garden fences to keep dogs in the garden and out of the paddock), as long as there is a second electrified fence preferably several feet in from the main fence. A less satisfactory option is to put an electric outrigger on the fence itself.

Chicken wire is sometimes used as part of a horse fence, especially on the lower half of the fence to keep small animals in or out. The stronger variety has been used successfully in foal paddocks as foals can bounce into it without injuring themselves.

Figure 8.7 A post and rail with chicken mesh fence Credit: Jane Myers

Figure 8.8 Diamond mesh fencing Credit: Diamond Mesh Fencing

Cyclone mesh fencing is good in small areas (due to its strength and expense) and for keeping escape artists in and persistent intruders out. A full height (about 3 m) cyclone fence will certainly prevent anything from leaving or entering the property but it tends to look industrial. This type of mesh needs to be fastened to something (as do many meshes) in order to keep its shape. Cyclone fence also comes in standard fence heights and can be fitted to a steel pipe or wooden rail at the top with either steel pipe or a very strong wire along the bottom edge.

Diamond mesh and Equi mesh fencing are mesh fences that are manufactured (by the same company) specifically for horses and are both an excellent type of fence but expensive. These types of fence are used in many studs because of their many safety features. The main ones are that horses, including foals, cannot get their hoof through the mesh and, because it is very closely woven, foals bounce off the fence if they hit it at speed. These meshes are manufactured in the USA and imported into Australia and New Zealand. They are usually fitted between posts with a top rail. The top rail is mainly for aesthetics and acts as a sighter for the horses. It does not require a bottom rail or wire. Once erected, the maintenance factor depends on the type of posts and top rail that are used.

Figure 8.9 Equi mesh fencing Credit: Diamond Mesh Fencing

Cable fences

Cable makes a very strong fence for horses. The cables can sometimes be obtained second-hand from electricity companies and if they are well strained (usually with 'turnbuckles' at either end) and these connections are covered, they are reasonably safe. A top rail of either timber or steel piping is usually required as the cables tend to sag easily, especially if horses are allowed to lean over them.

PVC/vinyl fences

PVC or similar material fencing comes in two main types. One type is comprised of PVC fence rails moulded over two strands of wire approximately 10 cm apart. These fence rails are supplied in long lengths (100 m rolls) that can be strained up to wooden, steel or poly pipe posts and can comprise the whole or part of a fence (i.e. three panels or just a top panel with wires beneath), forming a strong and flexible barrier. It is possible to get black as well as white rails. Another type of PVC fence comes in panels that look just like timber rails, with posts to match. PVC/vinyl fences have a long life span, are UV resistant and are low maintenance.

Wire fences

Wire fences are very popular fences for horses because they are relatively inexpensive, low maintenance and are reasonably safe if built properly. It must be remembered that

Figure 8.10 PVC fence rails — Credit: Flexafence Fencing

Figure 8.11 Vinyl post and rail fencing — Credit: Lee Bros. Fencing

any wire fence is more dangerous if it is loose. It is worth considering having strainers built in so that they can be tightened regularly and easily.

There are several types of wire available such as high tensile wire, soft wire, plastic coated wire or plastic wire.

High tensile galvanised wires such as 2.80 mm (11 gram) and 2.50 mm (12.5 gram) are strong wires that can be strained very tight and can withstand a lot of pressure. The problem with high tensile wire is that, if it does break, it can cause injuries as it snaps back or if a horse gets tangled in it.

Soft galvanised wires, such as 4.00 mm (8 gram) and 3.55 mm (9 gram) wires are more expensive than high tensile wire and require more frequent re-straining as they tend to loosen more quickly than high tensile wires. Some horse people prefer them, believing them to be less dangerous if a horse gets caught up in them.

Plastic-coated sighter wire is used for its strength and visibility. It can be strained tight and produces a good-looking fence. However, it is expensive compared to plain wire. It is also possible to buy plastic-coated wire that is able to be electrified.

Plastic 'wire' (Bayco) is also very visible but is not as strong as plastic-coated wire because it has no steel inner wire. When this is combined with electrified or non-electrified wires it is an effective fence.

Barbed wire is not a good fence material for horses. It does not stop the horse from leaning on it (only electric will do that) and it is expensive, very harmful to stock, dangerous to handle, and must never be electrified. It is at its most dangerous when it is not strained tightly. Horses can be successfully fenced without using barbed wire. If it is already on the property and it cannot be removed for whatever reason then it can be made safer with the use of an electric outrigger or better still, a simple electric fence can be installed several metres to the inside of it to keep horses well away (see 'Electric fences', below).

Electric fences

Electric fencing would have to be the most versatile of fencing options and can be the safest type of horse fencing. The use of electric fencing means that horses are much less likely to touch a fence, let alone lean on, rub on it or put their head through it. This, in

Figure 8.12 A wire fence with white sighter wire and short wooden droppers

Credit: Jane Myers

Figure 8.13 Electric sighter wire offset from timber post and rail

Credit: Gallagher Australia

turn, means that there is less chance of the horses damaging either themselves or the fences. Like people, horses vary in their response to a mild electric shock. Some are forever wary after the initial shock and others continue to challenge it if it is turned off for a period.

Electric fencing can be either permanent or temporary. It can make up the entire fence or complement other types of fence. Any handyman or woman can install and maintain it as electric tape or cord can be mended with scissors and a knot and it can be strained by hand. It is relatively inexpensive, and as it becomes more and more popular with horse owners electric fence manufacturers are responding by producing better-looking fences to cater especially for this market.

Portable electric fencing is an ideal fence for the sustainable management of a property because it allows you to quickly fence off new trees or areas of land that need protection for one reason or another.

Electric fencing can be used in many areas of a horse property with the exception of very small horse yards, training yards and arenas. It can be used on the perimeter fence (and is indeed recommended for this purpose) when it is used in conjunction with a solid type of fence, for example post and rail or posts and wire.

Electric fencing therefore works best in the following situations:

- as internal fencing (either permanent or temporary)
- to make unsafe fences safer (such as with certain types of mesh or wire fences)
- to protect certain types of fences (such as timber post and rail)
- to decrease the chance of a horse challenging the fence (such as with a perimeter fence).

The disadvantages of electric fencing are that it can be shorted out by grass that grows high enough to touch it or by sticks and branches that fall from trees and land across it. Regular checking will remedy this. Also it should be turned off on high fire risk days, especially if there is a chance of grass touching it. It usually takes horses some time to discover that the fence is no longer turned on so it is quite safe to do this from time to time. Electric fencing may be prohibited in urban areas (check with your local council). When it is permitted in these areas it must be clearly signed as such at frequent intervals along the fence.

The cheapest and quickest method of making an existing fence horse-proof (either to protect the horses from the fence or vice versa) is the addition of a single electrified wire, attached approximately 30–40 cm from the fence by outriggers. These outriggers can be in the form of plastic covered wire loops (see Figure 8.13), fibreglass rods, poly pipe lengths, hardwood brackets or wire outriggers with porcelain inserts (see Figure 8.14). Some can be attached to the wires of the fence, while others can be drilled into, nailed or bolted to the fence posts.

On an existing permanent fence one electric wire/tape/cord is usually enough for horses, fitted near the top, level with or higher than the top of the fence. If other types of animals also graze the paddock it is usually necessary to fit a second electric element at an effective height for these animals, which is usually lower than that for horses (unless you have miniature ponies and Friesian cows!).

Energisers

Electric fences require an energiser. This can be either a fixed mains operated unit, a fixed or portable solar/battery unit or a portable battery unit that uses either rechargeable or non-rechargeable batteries. A mains operated unit will need to be situated near a power

Figure 8.14 A wire outrigger with a porcelain insulator on a wire fence

Credit: Jane Myers

Figure 8.15 A solar energiser

Credit: Jane Myers

source so it is usually better to site it inside a building for security. Solar/battery or battery units are both portable. This makes them very useful but also susceptible to theft.

Energisers are available in various sizes so you need to work out how many kilometres of fence you will need to energise at a time. Most small properties can be adequately energised with one of the smaller units (5 km or less), depending on how many strands you plan to have and whether you will be electrifying all of the property at once or just one paddock at a time.

A small, battery operated, portable unit is very useful for strip grazing.

Wires, braids and tapes

There are various options for electric fencing ranging from the thinner types of galvanised steel wire to electric braid of different thicknesses and electric tapes of different thicknesses. Galvanised wire conducts better than braid and tape; it lasts longer but is not as visible. It is best used in permanent rather than temporary systems. Braids tend to last longer than tapes and tapes are the most visible option of the three.

The efficiency of an electrified wire depends mainly on its ability to maintain voltage over the length of the fence. Any leakage of current in large enough amounts can make the fence ineffective.

The effectiveness of a single electric wire depends on the electrical circuit made when the horse comes in contact with the wire. The horse may not get a 'kick' from the fence if the soil is very dry, or if the earthing system at the energiser is insufficient.

In drier areas it may be necessary to incorporate an earth-return wire with a live wire to get satisfactory results. Speak to your fencing supplier or manufacturer for details.

It is essential when designing your electric fence layout, that you incorporate a number of cut-out switches to isolate sections of fence for working on, or for fault-finding. This means that paddocks will be able to be switched off when not in use.

Figure 8.16 A wooden post and 40 mm tape fence Credit: Gallagher Australia

An electric fence tester is a must and, if you can afford it, a digital readout tester is far superior, as it will show you how much voltage is pulsing down your fence line. This is especially important when tracing partial short circuits. Some of these testers have an in-built arrow system, which shows you in which direction the fault is from the tester. Still others have a device for stopping and starting the electric fence from wherever you are along it so that you can get through the fence or repair it.

If you are using electric tape gates, set them up so that they are off when you unhook them to lead a horse through. This can save a potential accident if a horse steps on a 'gate' while being led through the opening. If a steel or timber gate is

Figure 8.17 PVC tread-in post with electric braid
Credit: Gallagher Australia

Figure 8.18 An electric fence tester Credit: Gallagher Australia

being used, an insulated wire can be run underground at the gateway or a tape gate can still be used to keep the horses back from the gate. This reduces the chance of head and hoof injuries from gates (see p. 170).

While it is true that electric fences are not generally dangerous to your horses, you must be careful when introducing a horse to electric fencing for the first time. Horses quickly learn to respect an electric fence but it is not a good idea to turn a group of inexperienced horses out all at the same time. A new horse which is brought onto the property, and which is not familiar with electric fences, should be turned out alone and allowed to discover the fence before being introduced to the other horses.

Most of the electric fence material manufacturers have excellent product brochures available which detail how to set up electric fencing on your property.

Posts

Permanent fence posts can be of timber (round, split or treated), poly pipe, concrete or steel.

Timber posts

Treated wooden posts are more expensive than untreated ones but generally outlast any non-treated posts (by as much as four times). Any wooden post will burn in a hot bush-fire, regardless of preservative treatment.

Untreated timber posts can be susceptible to rot, termites and fire. Some untreated timbers are far better than others to use for fencing and it is worth finding out what is

Figure 8.19 A treated wood post with electric braid
Credit: Gallagher Australia

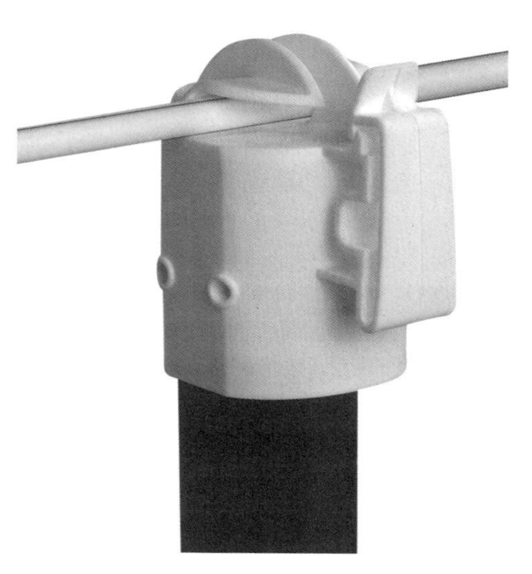

Figure 8.20 A plastic cap for steel posts
Credit: Gallagher Australia

the best type of hardwood and its life expectancy in your locality. Timber posts should be slanted or capped at the top to reduce rotting.

Recycled railway ties (sleepers) can also be used as posts and as they have been treated they last for a very long time. They give a wide flat area for fastening rails to which makes them ideal for post and rail fencing.

Steel posts (star pickets)

Horses and humans have been staked by steel posts, causing horrific injuries. Steel posts should always be used in conjunction with caps or sleeves that are purpose made for these posts. Capped or sleeved steel posts are best used as part of an electric fence system which keeps horses away from them. An electric stand-off wire or a second inner electric fence (using tread-ins and tape or braid) can be installed to keep horses away from the fence.

Figure 8.21 Fencing showing plastic sleeves over steel posts
Credit: Gallagher Australia

Steel posts are long lasting and are a boon in steep rocky areas or rocky areas such as gullies where it is difficult to get to with a tractor to knock in timber posts (typically the kinds of areas that horses either have restricted or no access to). These posts are also fire resistant so are recommended in areas where bushfires are common. To prolong the life of the wire they should be fastened to the posts with tie wire or other steel post fasteners rather

Figure 8.22 Portable electric fencing with pigtail tread-in post Credit: Gallagher Australia

than dragged through the post holes which removes the galvanised coating. Steel posts are also good for temporary fencing as they are light and easy to install with a hand-held rammer. To reduce costs and increase safety, place posts further apart and use droppers in the distances between them.

Poly pipe posts

These can be purchased as either recycled pipe from mines or other industries or new. They have many excellent qualities because they are chew, rot and insect resistant. They are not chemical treated; are lightweight; do not splinter; and have some flexibility if hit by a horse. They are also maintenance free. They can be rammed or dug into the ground in the same way as wooden posts. They can be cut to length with a chainsaw. No insulators are required for electric fencing and wire can either be threaded through holes drilled through the side of the post (remember to make them large enough so that the wires do not drag) or through fasteners that have been attached by tech screws. With these posts a hole should be drilled near the bottom so that rainwater can escape and does not short out any electric wires threaded through them.

Electric fence posts

For a permanent electric fence any of the posts described above can be used. Posts other than poly pipe will need insulators attached to them. There are also several kinds of kind of post that are made specifically for electric fence systems.

Tread-in posts

These are made of UV stabilised PVC or steel with a PVC covered top pig tail loop. These posts are for portable fence systems such as when strip or block grazing a paddock.

Figure 8.23 Electric fencing with fibreglass post Credit: Jane Myers

Figure 8.24 Stainless steel clip on a fibreglass post Credit: Jane Myers

Figure 8.25 Equifence^{TM} and Insultimber^{TM} posts Credit: Gallagher Australia

Fibreglass posts

These are electric fencing posts that are self-insulated. Wire is attached to them with stainless steel clips. The fibreglass posts can bend without breaking and are ultraviolet stabilised. They are available in a range of lengths with 10 mm diameter posts for temporary electric fencing and 13 mm diameter posts for permanent electric fences.

Timber posts

A dense eucalypt commonly called Ironbark is used for posts and droppers/battens for electric fences. This is marketed under the trademark Insultimber^{TM}. They have good insulation characteristics and do not need to have insulators attached when used in electric fencing.

Droppers/battens

Using droppers or battens to space and support the wires means that a wire fence can be constructed using fewer posts. These droppers or battens can either be full length and

Figure 8.26 Fence showing wooden droppers, wooden posts, sighter wire and electric wire Credit: The Horse Shed Shop

Figure 8.27 Electric fence showing poly pipe droppers Credit: Jane Myers

rest on the ground, which gives the wires more support, or shorter lengths that simply space the wires above ground level. As they are less expensive than posts both to buy and to fit, their use will reduce the total cost of the fence system. Fence wire is usually attached to them by either being threaded through or fastened with soft wire to pre cut holes or grooves.

Wooden droppers

Sawn hardwood or treated pine droppers are available with a number of options. They can be bought as plain board or with holes or slots pre-cut. Ironbark droppers (Insultimber™) are commonly used in electric fencing systems as they are self-insulated.

Steel droppers

Pre-formed light-gauge steel droppers can be used for the same applications as steel star posts. They can also form part of a wooden post and wire fence.

Poly pipe droppers

Droppers can be made from poly pipe – they are easy to make and to erect and are cost effective. Holes can be drilled to thread the wire through, but preferably grooves should be cut into the pipe to hold the wire or soft wire should be used to attach the fence wire after it has been strained.

Erecting fences

When erecting fences, there are several options available to you.

- *You can buy the materials and do it yourself.* This will be the cheapest option, but is only any good if you have the time, inclination and skills. Fencing can be rewarding, but if you have no previous experience try to find a short TAFE/tech college course that will teach you some skills.
- *You can get a local fencer to put the posts in for you and do the rest yourself.* This is a good option as the heavy work is done for you (without the right machinery this is very difficult to do alone).

Figure 8.28 Timber and wire fence Credit: Jane Myers

- *You can contract a local fencer to erect them.* A fencer may also have better buying power when it comes to obtaining materials. It is a good idea to compare prices and decide whether it is better if you supply the materials or they do.
- *You can get a commercial fencing company to erect them.* This will be the most expensive option. However, this way you will be able to get a guarantee for both the workmanship and the materials from the same place.

If you decide to have a go at putting up your own fences there are certain guidelines that you can follow to make it easier. Also a certain amount of equipment will be required (see Chaper 11).

Posts can be set in an oversize hole which is then backfilled or they can be driven in using a post rammer on the back of a tractor. The diameter of a timber post has very little effect on its stability. However, a post that is driven has been shown to be one and a half times as firm as a post set in an over-size hole that is then backfilled. Posts can be concreted into an oversized hole but only concrete the lower end and then backfill with tamped earth. Posts that are concreted for the full depth of the hole can break on impact due to being less flexible.

A hole can be dug by hand or with an auger. These can be hired as a separate piece of equipment or as an attachment for a tractor. Posts are set 1.5–4 m (6–12 ft) apart depend-

Figure 8.31 End assembly Credit: Jane Myers

Figure 8.29 (a) and (b) Stays Credit: Jane Myers

ing on the kind of fence. Post and rail will be closer together; a simple wire fence can have far apart posts if droppers are also used. Use a carpenter's building line to get your lines straight, this is not just for neatness – otherwise the fence will move when strained up.

Aim to put about 30 per cent of the post underground. If the soil is heavy it may be possible to reduce this proportion. The deeper set the post is underground, the more resistance it will have to being pushed over.

End assemblies or stays are required at the ends of a wire fence line and are the foundations of a fence. Steel end assemblies are available and they are preferred in areas of high fire risk. They are more expensive than wooden ones but are simpler to erect.

There are a few types of timber end assemblies, but the most successful is the horizontal stay 'H' assembly. The end assembly must be 100 per cent effective, or else the fence will fall over. End assemblies must be fitted at each end of the fence line, and each side of a gate. Ideally the top stay rail should be 2.25 to 2.5 times the height of the posts, with the diagonal wires tight and around the base of the post at the furthest end of the fence line.

Stays are easier to erect than end assemblies and they work well if erected properly. On a small horse property the fence lines are usually quite short so a well-made stay is usually more than adequate.

Use a wire spinner on a trailer when running out plain wire. Tie one end of the wire to one end post and then drive to the other end post. This minimises the damage to the zinc galvanising and lengthens the life of the wire. Wire should not be dragged over the ground or through holes in posts or droppers. Not only is zinc scraped off, but also the holes harbour water, increasing the chances of corrosion. Holes are funnels for heat during fires and will promote burning. Strain the wire to the recommended load using a load-measuring strainer. Once strained, the wire can be stapled to the fence posts or tied in or clipped in to other kinds of posts/droppers. Wire tensioners can be used so that the wires can be frequently tightened.

Figure 8.30 A wire tensioner Credit: Jane Myers

When stapling the wire, use a staple driver to save your thumbs. Choose the right staple for the job (your fencing salesperson will advise on this). Angle the staples down-

Figure 8.31 A close gate fitting

Credit: Jane Myers

wards and at 45° to the vertical. Stagger them as you work down the post to avoid splitting. Do not knock staples all the way in otherwise the fence cannot be tightened later on.

If possible fix the wires to the northern or western side of the post. Then if there is a grass fire and the post burns, the northerly wind will tend to fan the heat away from the wire, increasing its chance to be salvaged.

Gates and gateways

Gates and gateways are potentially the most dangerous section of a fence. This is because this area is where the horse may spend time waiting to be let in for feed or where the horse either takes itself through the opening (in the case of gates left open) or the horse is led through. Common injuries associated with gates and gateways include those listed below.

- Injuries (sometimes fatal) to the face, neck or legs when a horse gets trapped between the gate and the post when rubbing or pawing at the gate.
- Similar injuries occur when two gates are used. The gates cause a nutcracker action if the horse traps its head or legs.
- Injuries caused by horses jumping or attempting to jump gates because they are 'cornered' by other horses or they are trying to get towards something.
- Injuries to hoofs and legs if the horse gets a foot trapped in the gate by pawing at the gate.
- Injuries to the body from protruding gate fasteners as the horse is led through the gate or slams into the gate when galloping around a paddock. These can also cause injuries to people if they get pushed into the gatepost by a horse. Horses can also get a leg strap of a rug hooked up onto a gate fastener as it is led through the gate.
- Horses can get trapped between the gate and the post when the gate only opens to the side they are on.
- A gate can comes off its hinges and fall onto a horse when the horse rubs its head on it.

The incidence of these injuries can be reduced or eliminated by careful attention to safety aspects when you construct fences and/or gates, as detailed below.

- Fit gates flush to posts, and preferably use square-edged rather than rounded-edged gates that leave no gap for the horse to get its head into (see Figure 8.31).
- Fit double gates flush to one another with a fastener that does not allow the horse to get its head between them (see Figure 8.33).

Figure 8.32 A gate fastener

Credit: The Horse Shed Shop

Figure 8.33 A double gate fastener

Credit: Jane Myers

- Avoid having gates fitted right into the corner of a paddock (aim for at least 3 m to 6 m in from the corner). Introduce horses carefully to other horses before turning them out together (see p. 16).
- Ensure gates are either closely meshed so that a horse cannot get a foot through, or that gaps are large enough for a horse to get its foot out easily. Avoid using gates that have diagonal bars as a horse can trap a leg between this and a horizontal bar (see Figure 8.34).
- In paddocks where the gate is situated in the corner, fit an electric tape across the corner with a hook. This keeps horses away from the gate when they are in the paddock and makes it easier and safer for people and horses when putting a horse in the paddock and taking one out. It also reduces the chance of a horse getting cornered and jumping the gate as a result.
- Make sure gate fasteners do not protrude at all and are tamper proof from horses. This is difficult to achieve as many fasteners on the market are dangerous for horses. Gate fasteners can be made safer by recessing them into the post. As many gateposts are timber this is usually possible. If not, fit the fastener so that it does not protrude where a horse or handler can get caught up on it.
- Paddock and yard gates should swing freely and open both ways.
- Make sure the gate cannot be lifted off its hinges if it is rubbed by a horse.

A gateway should be at least 1.2 m wide for horses. However, it is usually better to make a gate at least 3 m wide so that it can be used for machinery as well.

Gates can be made of various materials with some of the most common being pipe, timber, electric tape and wire.

Pipe gates are the most common type of gate. However, some are better than others for horses. A common type of pipe gate has rounded corners and is meshed with large gapped mesh. The rounded corners are dangerous unless fitted very close to the post and the mesh is often too large, allowing hoofs to get caught. A better type of pipe gate is either made up of horizontal and two vertical bars (no diagonal bars) and no mesh or is meshed with 10 cm mesh or smaller.

Timber gates are usually only used with timber fences as they are expensive and due to their weight they tend to sag. Heavy gates can either be rested on a stump or stone at

Figure 8.34 A close meshed gate with PVC rails attached to timber post

Credit: Jane Myers

Figure 8.35 A horse safe gate

Credit: The Horse Shed Shop

the other end from the hinge when closed, or be fitted with a wheel at this end to help support it and help with opening and closing. Timber gates would usually have to be made to measure as they are not commercially made due to their expense and therefore reduced demand.

Pipe and timber gates should never have diagonal bars (unless they are also meshed) as they can trap a hoof in the acute angle caused by a horizontal bar meeting the diagonal bar.

A 'gate' can also be made from electric fence tape or cord. These can be used as the gate itself on internal electric fences or to make a solid gate safer.

9

Trees and plants

Both in Australia and New Zealand trees are cleared on a large scale every year. In fact many more trees are cleared every year than are planted. Australia has one of the highest rates of tree clearing in the Western world – much of this is for subdivisions and housing.

Establishing and caring for trees and plants is a very important part of property management. Trees and plants provide many vital functions both on a micro and macro scale. Aim for approximately 30 per cent of your land to be vegetated with trees.

The benefits of trees and plants

With careful selection and placement, trees and plants can do all of the following:

- provide shade and shelter from sun, wind and dust
- slow down fire
- give privacy from neighbours
- provide habitat for wild animals including birdlife
- add value to the property by creating an attractive environment in which to live
- modify the climate of a property and in turn other plants, animals and people on the property will all benefit
- provide feed for stock
- give natural pest protection
- are an aid to drainage and erosion control
- reduce the effect of global warming.

Trees and plants as habitat

Trees and plants are habitat for numerous species of insects, mammals and birds. There are many benefits of attracting various species, some examples are:

Figure 9.1 Revegetation Credit: Jane Myers

Figure 9.2 A shade tree Credit: Jane Myers

- certain birds are primarily insect eaters and help to control pests such as mosquitoes, flies, caterpillars, grasshoppers and aphids
- species of native beetles, spiders, centipedes, native wasps, bees and other insects all have a ecological niche and play an important part in maintaining an ecosystem
- sugar gliders (in Australia) mainly eat insects, including those that damage plants
- small insectivorous bats eat many insects, such as mosquitoes at night-time, sometimes by the thousand.
- frogs also eat some of the less desirable insects.

Find out which birds and other beneficial animals or insects migrate to or live permanently in your area and encourage them by planting trees and plants that attract them. Many Australian and New Zealand native trees and bushes are bird and bat attracting. You may have to provide nesting boxes, as it is only very old trees that have

Figure 9.3 Land for wildlife Credit: Jane Myers

the hollows that some birds need for their nests.

Leave old dead trees where they fall as they provide habitat for wildlife. Rocks provide habitat for frogs. Make sure these areas are fenced off and not a danger to your horses.

Remnant vegetation

Remnant vegetation is any area of bushland that has either not been cleared or is regrowth from previously cleared land. Often these areas are seen as a nuisance on a property and are earmarked to be 'tidied up'. Instead of being regarded as a problem they should be seen as a valuable asset. These areas provide habitat for flora and fauna, help to keep salinity at bay and provide windbreaks. The animals and insects that live here help to control problem insects on the remainder of the property.

This vegetation is sensitive and must be fenced off from domestic grazing animals to protect it. If possible these areas should link with other remnant vegetation both on your property and outside the property to maintain wildlife corridors. Any fencing should allow native animals to pass through.

Any existing weeds in these areas must be tackled, the problem being that because they become unproductive in terms of pasture they tend to get forgotten about. As well as causing problems for wildlife, noxious weeds will also infest the pastures.

Trees for shelter and fire protection

Trees can provide very effective shelter for paddocks and yards. They should either be planted outside the paddocks or yards so that they are protected from horses or fenced for protection if they are planted within a horse area.

Deciduous trees can be useful if planted on the north side of yards so that they provide summer shade, and allow these areas to dry out in the winter when they lose their leaves. Some deciduous trees can also be used as fodder trees such as willows, chestnuts and poplars. Other trees that are good for shelter near yards are dense bushy trees. Choose species that are fire resistant and are not poisonous to horses.

Windbreaks and firebreaks

Trees as windbreaks help to stop moisture loss (by approximately 20–30 per cent) and soil loss to wind. They need to be across the line of the prevailing winds to be effective. Plan shelterbelts that will benefit areas such as yards and paddocks as well as the house and other buildings. Double fences and laneways between paddocks provide an excellent area to plant windbreaks and these have the added advantage of helping the horses to see the fence as a physical barrier.

Figure 9.4 A new windbreak Credit: Jane Myers

Figure 9.5 A well-established windbreak Credit: Jane Myers

A good windbreak slows the wind down protecting everything on its lee side. Windbreaks should be constructed so that they slow the wind speed by 60 per cent to be effective; solid windbreaks do not slow the wind as much because the gusts hit them and jump over them whereas permeable ones slow the wind as it goes through. A certain amount of wind must be able to pass through a windbreak or else the effect manifests as eddy currents on the lee side. A good windbreak will extend its effect 20 times its height downwind and 5 times its height upwind. The trees need to be stepped by planting in at least three rows with the tallest trees in the middle. This is achieved by planting different species that have different heights at maturity.

Windbreaks should also be designed with fire in mind (see p. 59). They should not be too close to buildings or yards. For a shelterbelt of trees to function correctly, it should be at least 20–25 times as long as it is high. It should also be made up of flame-resistant varieties of trees.

Flame-resistant trees and plants tend to have certain characteristics in common. These are:

- they are slow growing (e.g. deciduous trees)
- they have a high salt content (such as mangroves and salt bushes)

Figure 9.6 An effective windbreak should be able to be penetrated by the wind, which reduces the wind speed and velocity. Credit: Jane Myers

Figure 9.7. A solid windbreak creates air turbulance on the leeward side, which is not desirable.
Credit: Jane Myers

- they do not collect litter within the canopy or on the bark (they have smooth barks rather than stringy or paper barks)
- they contain high moisture levels in their leaves (rainforest plants, cacti, succulents and most fruit trees)
- they contain low volatile oils.

Trees to avoid are pines, conifers and most sclerophyll trees such as gums, melaleuca and callistemon.

Even fire retardant trees and bushes will burn eventually. However, they do offer some protection. They reduce the wind speed and this reduces the speed and intensity of the fire. Trees absorb the radiant heat, protecting animals and buildings on the lee side. They also catch airborne burning material, stopping it from flying into other areas and starting spot fires.

In particularly high-risk areas where the windbreak is too close to buildings, remove the middle storey of plants to reduce the risk of bushes igniting the tree canopy. The base

Figure 9.8 A tree-lined driveway Credit: Jane Myers

of trees and bushes needs to be kept clear as any matter collected there (for example fallen twigs, dead leaves and dry foliage) counts as fuel.

Hedges of plants such as privet, boobialla and pittosporum, planted no closer to buildings or yards than 6–10 m, can act as radiation shields. Most native shrubs, however, are too flammable to put close to buildings, and neatly trimmed cypress hedges are extremely hazardous because of the dead clippings that accumulate in the centre of them.

Every area differs in what can be grown and what is available. Local conservation departments will be happy to provide you with the names of useful shelterbelt and fire-break trees for your district.

Trees and plants for fodder

In some parts of the world trees and plants (other than grass) sustain animals. Grasses generally have shallow roots and cannot survive in very harsh conditions. There is a growing interest in feeding domestic animals with fodder trees in Western countries due to the success of fodder trees in countries with harsh or even desert conditions.

Most fodder trees share some similar characteristics such as:

- they have deep roots so that they can access water from deep below the surface and provide fodder even in very dry conditions
- many fodder trees provide more feed than pasture
- many are legumes (such as tagasaste and acacias) therefore they fix nitrogen in the soil eliminating the need for nitrogenous fertiliser
- many have high protein yields
- many fodder trees are also good windbreak and fire-resistant trees.

When sourcing fodder trees, choose local native species whenever possible. You also need to take into account your climate and soil conditions.

Fodder trees and bushes need to be pruned or grazed regularly so that they do not grow too tall. Aim for no more than 2 m in height so that both you and your animals can reach the top. Do not plant them as boundary plants as they need to be able to be accessed from both sides, therefore laneways and driveways are best. They can also be

Figure 9.9 Fenced willows

Credit: Jane Myers

planted around arenas and yards. You may want to plant them as an orchard of trees and allow the horses in for periods of grazing. A variety of fodder plants would be a good idea in this situation so that horses can browse as they would naturally.

The plants need to be alternately grazed and rested, just as you do with pasture. They will need to be fenced off from stock for at least the first year so that they can get established.

In New Zealand farmers use tagasaste, saltbush and willow to feed stock. Australia has access to many fodder trees that are imported, as well as natives such as acacias, casuarinas and saltbushes.

Some plant species that can be used for fodder are listed below.

- *Willows*, especially weeping willow, these trees are fast growing, yield 17 per cent protein. They can be cut or grazed right back, they are also drought tolerant and frost resistant.
- *Poplars* suit conditions similar to willows, they are deciduous fast growing trees. Plant them well spaced out. These trees tolerate dry periods.
- *Carob*, a slow growing tree, produces pods that are fed to stock (seeds are 21 per cent protein). They are a legume and will grow in a variety of soils and climates, but they are not frost tolerant. High yielding (50 tonnes of pods per hectare per year).
- *Honey locust*, grown for a high yield of pods, frost tolerant it can survive in dry climates and will have rapid growth in areas of significant rainfall.
- *Tagasaste* is an evergreen high protein (23 per cent) fast growing tree that reaches its full potential in three to four years. Once mature it is frost tolerant and can be completely defoliated. It is a legume that likes free draining soils and is drought tolerant. It is fire resistant if kept as a low hedge. Tagasaste should be controlled so that does not invade native vegetation, you should aim to prevent flowering by pruning or grazing.

Figure 9.10 A fodder tree Credit: Jane Myers

- *Leucaena* is a tall legume, high in protein (18 per cent) shrub that suits a wide variety of soils including poor soil. It is sensitive to frost but can recover from complete defoliation
- *Wattles* (acacias) are legumes, good species are golden wreath wattle and mulga. They suit low rainfall areas and will grow in most soils including clay soils.
- *Saltbush* and *blue bush* are high protein quality feed with high yield. They are salt and poor soil tolerant.

Poisonous trees and plants

Many varieties of trees and plants can cause poisoning in horses. Some are more dangerous than others, some are more likely to be eaten than others, and some are only poisonous at certain times of the year/or in certain climate conditions. Some trees and plants have to be eaten in large quantities before affecting the horse, while others are poisonous in very small amounts. Also, horses differ in whether they will eat poisonous plants (some will, some won't). This can vary depending on how hungry/greedy the horse is. A hungry horse is also likely to be poisoned more quickly because of its more rapid absorption of toxins. What the horse learned to eat or avoid when very young also determines consumption. Foals learn what to eat and what to avoid from their dam. This means that horses that are exposed to plants that they are unfamiliar with may eat them when other horses that are accustomed to them would not.

It is not always easy to identify which plants may cause poisoning or disorders in horses. Without botanical knowledge, it is impossible to distinguish between a poisonous and a harmless plant. There is no characteristic of taste, smell or appearance which indicates which is which, and no simple method of identifying the poisonous ones, or when poisoning is likely to occur. Fortunately horses are very selective and therefore cases of plant poisoning are not common unless horses are hungry.

The following precautions must always be observed to minimise risk.

- Learn to recognise poisonous plants and be suspicious of unfamiliar plants to which horses might have access.
- Find out how such plants can be removed.
- Recognise the conditions when some plants are at their most dangerous.
- Do not allow horses access to garden waste or poisonous trees near fence lines.
- Seek veterinary advice as soon as poisoning is suspected.

Horses suffering from plant poisoning may show a wide range of symptoms including:

- sudden death or death within a few days
- nervous diseases – e.g. staggers
- photosensitisation – excessive sunburn on white skin areas
- liver damage
- kidney damage
- skin disorders
- heart and lung disorders
- intestinal disorders.

When a horse's condition indicates that plant poisoning may have occurred it is important to check:

- what signs of illness the horse is showing
- what plants the horse has been feeding on
- what plants are available in its grazing area
- whether the horse can reach native or ornamental plants over a garden, roadside or windbreak fence
- whether these plants, and their toxic properties can be identified
- whether neighbours or passers-by may have fed or dumped anything over the fence.

Seek professional veterinary help immediately and provide the veterinarian with as many answers as possible to the above questions.

Prevention is better than cure. Owners should be aware of the possibility of poisoning and at least be able to identify the most common poisonous plants both in their local area and in general so that any new noxious weeds springing up in an area can be quickly identified and taken care of. Find out if your local council has a Weeds Officer and buy or borrow a good well-illustrated book that will help you to identify weeds and poisonous plants (see Chapter 12 for where you can find out more about poisonous plants).

Trees and plants for pest protection

As well as providing habitat for birds and other animals that eat pests, plants can also be used for pest protection. Certain plants will repel insects such as thyme, rosemary, lavender, lemon verbena, star anise, tansy, chamomile, mint, elder, basil and bay. Repellent bushes include citronella, lemon scented geraniums and pelargonium (see Chapter 12 for recommended reading on pest repellent plants).

Buying and planting

Once you have decided what trees and plants you want you will need to find a good supplier. If you are vegetating or revegetating your property, bit by bit, it may be possible to get what you need on an ad hoc basis, looking for good buys at nurseries and markets and even auctions. If you are planning to buy in bulk you may need to try the larger nurseries, some of which are run by local councils or other government bodies. Your local Landcare officer is another excellent contact as it may be possible to get subsidised or even free trees when you are vegetating or revegetating a property, particularly if it involves a waterway.

Make sure that the plants that you buy are healthy specimens. Tube stock – plants or trees in narrow tubes – are usually much cheaper to buy and if they are purchased at the right time (before they outgrow the tube) a high success rate will be achieved. Tubes allow the roots to grow in the correct way for plant development. If you are buying large amounts this is a very good way of buying trees. Plan all your purchases well in advance as you may need to order if buying in bulk.

Before you plant you need to prepare the area for the plants. Compacted soil will need to be ripped by a tractor with a ripper implement so that young roots can infiltrate; waterlogged soil will need to be mounded so that the new plants do not drown. Weeds will need to be controlled before new plants are planted. Once the new trees are in the ground, weeds can be further controlled by mulching.

The time of year can make a big difference to the success of your planting. Aim to plant before the wettest time of year so that the plants can take advantage of the rain when it comes and are already well established before any prolonged dry periods.

If planting into undisturbed soil make sure that any holes for plants are significantly larger than the pot size. This hole can be backfilled with loose soil once the plant is in the hole. Do not use a post hole digger (borer) in clay soils for plant holes as this tends to seal the edges of the hole making it difficult for young plants to penetrate further with their roots.

Putting a handful of 'blood and bone' in the bottom of each plant hole will give the plants a good start. If your soil is sandy and dries out quickly put some better quality soil mixed with a teaspoon of Rainsaver Watercrystals™ in each hole. These water crystals swell enormously when wet, holding water in the soil and releasing it slowly to the plant roots. The crystals continue to work for up to five years.

Plant large trees 4–5 m apart, large bushes/small trees 3–4 m apart and shrubs 2–3 m apart and combine plants that have similar growing needs and performance together.

Protection and care

Trees and plants need protection from various quarters if they are to survive and thrive. A healthy tree or plant will be able to resist pests and diseases. However, it is important that they are protected from being abused by livestock. Even fodder trees need to be allowed to become established and then given regular breaks: they will die out if the browsing of them is not controlled.

If you have mature plants and trees on your property they will need to be fenced off to protect them from livestock. Horses will eat the bark of trees – and ringbarked trees cannot survive. In addition, large grazing animals such as horses and cattle compact the soil around the base of the tree when they use the tree for shade, chewing and rubbing. This is why, even though scattered trees look nice in a paddock, they are difficult to manage because each individual tree will need protection.

Trees can either be fenced off in groups or singly. The latter is an expensive solution so if this is not possible you have a few other options. The trunks of trees can be protected by wrapping them individually in wire mesh, tin, corrugated iron etc. However, this does not protect the base of the tree from compaction. Mulching around the base will give some protection and cushioning from hoofs but more effective is to place a ring of tyres around the base, as most horses will not step into or over tyres willingly. Make the ring a generous size to protect the roots. This is especially important for trees with a spreading canopy as these tend to have spreading roots.

Trees that are fenced off in groups will be healthier as trees rely on each other in many ways. Any gaps between the trees can be allowed to revegetate or planted with species that you want on the property such as fodder trees, bird-attracting trees and so on.

Figure 9.11 A protected tree Credit: Jane Myers

Young trees and plants need to also be protected from rabbits and other grazing animals. Rabbits will eat young trees and plants before they get a chance to grow. Putting a plastic or cardboard 'tree saver' around each plant will protect the plant from rabbits. The tree saver also creates a mini greenhouse around the plant which gives it added protection. Tree savers will not protect plants from horses as they will pull the plant out from the top if they want to eat it. Only fencing of either a temporary or permanent variety will protect trees from foraging horses.

Young trees and plants are more susceptible to variations in temperature than more mature plants. However, an unusually heavy frost can affect and even kill plants that are mature. Some parts of Australia as well as New Zealand can have very cold winters as well as hot summers. At certain times of the year and at certain (higher) altitudes there can be a huge difference in the daytime and night-time temperatures. Certain plants that are not native to the area will need to be protected against the elements if they are to survive to maturity. Mulching around the base of the plant (but not touching the stem or trunk) will help to regulate temperature extremes. Plants should receive reduced watering and no fertiliser at the end of the growing season so that they do not grow new leaves which will be more prone to frost damage.

Mulching

Mulching around the base of plants suppresses weeds, acts as a slow release fertiliser for plants (depending on the mulch type), provides an environment for plant friendly insects and reduces evaporation.

Many materials can be used as mulch such as grass clippings, paper, straw/hay or woodchips. Any organic matter will make mulch (some better than others). Grass clippings are best if mixed with rougher material such as twigs and leaves as they tend to clump. Fresh manure should not be used as once it dries it tends to repel water and it can be too 'strong', especially for some native plants. Composted manure is good. However, this should be regarded as a fertiliser and covered with another type of mulch so that it does not dry out. Newspaper, cardboard or carpet placed under another form of mulch will reduce evaporation even more (by up to 70 per cent). In areas close to buildings, gravel can be used as a mulch to create a firebreak.

Figure 9.12 Tree saver made from cardboard Credit: Jane Myers

One of the best forms of mulch is leaf or bark (such as pine bark) mulch (not pine woodchips which have no leaf content and are not much good). Tree prunings which contain a high percentage of leaf material are usually readily available from businesses that advertise tree lopping. These become a by-product when the business has pruned trees for a client and the client does not want the mulch. Also the local council often has large amounts of this to either sell cheaply or give away (this depends on your council). If the mulch is to be used where horses will have access to it, check that it does not contain poisonous plant material.

Be generous with the mulch and pile it higher on the outer than the inner rim. This will help to protect the base of the plant. Mulches, especially grass clippings, should not touch the base of the plant as they can cause it to rot.

10

Manure management

The valuable by-product, manure, is often considered a waste or, at best, a nuisance to dispose of. Yet most horse properties, including small ones, can use this manure to their advantage, it just takes a little time and planning.

In addition, in urban areas you will come under the close scrutiny of your neighbours, so it is a good idea to keep on the right side of them by demonstrating that you are responsible and are managing manure properly.

Manure management options

Several options are open to you to manage any manure that is collected from stables, yards and paddocks. These options include:

- collected manure can be composted to spread later on paddocks, gardens or to be sold
- collected manure can be spread on paddocks immediately, sold or given away without being composted
- manure can be put in special garbage containers from a waste disposal company and removed regularly from the property.

The third option should be a last resort as it is an expensive solution and can only really be justified in situations where the amount of horses kept on a property is far more than the property can support with pasture. A typical example would be an inner city riding school or racing stable where there will be very little or no pasture and many horses kept stabled or yarded in a relatively small area. Even on this kind of property it still may be possible to compost and then sell the manure, in this way turning a liability into an asset. In an urban or city area, compost is an even more valuable commodity. If this is not possible, try to find a company that will either buy or remove for free your manure on a daily or weekly basis.

Figure 10.1 Manure for sale

Credit: Jane Myers

Effects of types of bedding

Whatever you chose to do with your manure, the type of bedding that you use will affect its usefulness. Too much bedding, or certain types of bedding, can slow down the rate of decomposition. This is important if you are going to compost it yourself or give or sell it to someone else to compost. Wood products are high in carbon and need to be balanced with high nitrogen products (i.e. manure) to work. Grass clippings, leaves and chicken manure are also good additives to balance too many wood products. Fine sawdust can work well because of its small particle size. Straw and shredded paper are better if they are chopped (with a mulching machine) before adding to the pile to speed up decomposition. Aim to reduce bedding use for better composting. Using rubber mats vastly reduces the volume of bedding required, making composting much easier. By changing your bedding type or by using rubber mats the manure becomes a more usable and saleable commodity.

Manure accumulation and composition

A stabled or yarded 500 kg horse can generate 8–10 tonnes of manure a year. Composition of this material varies depending on the type and quantity of bedding used, the age and function of the animal, the type of feed and how the manure is stored. Typically, a tonne of fresh horse manure, with bedding, would have a nutrient composition of about 5 kg of nitrogen (as N), 2 kg of phosphorus (as P205), and about 6 kg of potassium (as K20). About half of these nutrients may be available to a crop during a growing season with a spring application. Part of the remaining nutrients will provide fertiliser value in subsequent years.

Manure also contains other valuable trace elements. In addition to providing nutrients, manure improves soil texture and soil moisture-holding characteristics, thereby reducing the need for irrigation.

Manure storage

A single horse will produce 300 cubic centimetres (cm^3) of manure a day. Bedding can easily bring the total volume of material that must be managed each day to 500 cm^3 per animal, depending on the type of bedding used. Using rubber stall mats reduces the amount of bedding needed to a minimum (see p. 140).

Provision must be made for proper handling and storage along with a plan for effective utilisation of manure. Decomposition of manure starts as soon as it is passed. Decomposition rates depend on handling and storage methods. Stored horse manure should be kept compact and moist to prevent excessive loss of nutrients. Manure left in a loose heap loses nitrogen rapidly to the atmosphere in the form of ammonia, is available to flies for laying eggs and can end up contributing to nutrients entering the waterways via run-off. Because you will need to build storage facilities for your manure, you may as well consider composting it.

Composting manure

Composting manure is the best way of turning a liability into an asset. Composting is a method of speeding up the process of decomposing that occurs naturally to everything organic. By providing the correct conditions, the micro-organisms and bacteria are able to work to their full extent.

Benefits of composting manure and bedding

- Compost makes nutrients such as phosphorus more available to plants as it changes the organic matter into substances that more readily form humus in the soil. Nitrogen in particular, instead of being highly soluble (and therefore more able to be washed into waterways) as it is in fresh manure, is converted to a more stable form that is more slowly released. This slow releasing of nutrients means that nutrients are available to plants (albeit in decreasing amounts) in the seasons after application. With regular topping up of composted manure the need for other fertiliser is very low or eliminated altogether.
- As well as adding organic matter to the soil, compost improves aeration and water retention. This means that in sandy soils compost helps the sand to hold water (compost can hold almost twice its own weight in water), and in clay soils compost loosens the packed clay by opening up pore spaces which allows water and air into the soil.
- When soils are able to hold water for longer they do not dry out as quickly in a dry period and therefore do not erode as quickly from the wind.
- Improved soil feeds and encourages earthworms.
- Compost gradually changes the soil pH in soils that are either too acidic or too alkaline.
- The high temperatures achieved when composting kills weed seeds, parasites and pathogens such as bacteria, viruses and fly eggs and larvae.
- In properly enclosed compost systems the breeding ground for flies is also reduced (compared to an open manure pile), further reducing the fly problem.

Figure 10.2 Compost and spreader Credit: Jane Myers

- Odours are also reduced especially in a covered bin.
- Composted manure has roughly 40–60 per cent less volume and weight than non-composted manure.

Once manure is composted you are left with a material that is high in organic matter and is similar to potting soil. This material is valuable and should be used on the property for improving soils in paddocks and gardens. Of course it also has a sale value – but why sell such a valuable commodity unless you already have excellent soils? Can you imagein a farmer selling manure? Of course not – it's all used on the land.

To compost manure, you must pile it properly, keep it moist, and either turn it over several times for one or two months or establish a system of aeration. Various techniques can improve and hasten the composting process. Processing methods can be kept quite simple, or be quite sophisticated, depending on the desired condition of the end product and the time needed to complete the composting process.

The amount of space required to compost your manure depends on how many horses you have and how they are managed. Other considerations are: Do you have access to machinery or will you be moving it around by hand?

A 10 m × 10 m pad will comfortably house three compost bins with room to move machinery. This area is large enough to cope with the manure from approximately five horses that are kept confined on a regular basis. Three bins are ideal so that one can be filling, another composting and the other full of finished composted material that is available to use. A smaller area will be fine for properties where less manure is collected (i.e. the horses are paddocked for much of the time) and/or machinery is not used in the process. A good idea is to start with a smaller area and add to it if the need arises later on. Position the pad with this in mind.

The ground surface should be non-permeable to prevent leakage of manure into the soil, which can then get into the waterways. If possible, a concrete pad should be laid. This is even more important if machinery will be used to shift the compost, otherwise the area will quickly become waterlogged and smelly.

Grade the surrounding area to keep surface water from running over or through the manure and into streams or other surface waters. Covering the compost bins will also prevent liquid from leaking into the storage area and it reduces flies.

A buffer zone of vegetation is required between the compost area and any waterway. It needs to be positioned as far away from a watercourse as possible. Check with your local council about regulations for your area. Try to locate compost bins so that filling and empting them is convenient. It may be possible, for instance, to locate a compost bin below the level of the stables so that manure can be tipped into the bin without having to fork it out of the barrow (see Figure 10.2).

Bays should be built for efficient composting as unless a pile is compact it does not reach the required temperature to compost. A pile should be at least 1 m high to reach high enough temperatures and 1.5–2 m^2 in area. Three or four sides on a compost bin can be constructed from various materials including bricks, old tyres, bales of hay, concrete blocks, steel sheets, timber, concrete and so on.

Air and water are required for decomposition to take place. The pile should either be turned at regular intervals or you should insert some form of perforated piping into the pile to get air into the centre. The compost bin can be constructed with slatted sides for more aeration. The pile should be kept damp but not too wet (the consistency of a wrung-out sponge). Covering the pile will help it to stay moist without allowing rain-water to get in and make the pile too wet. Once the bin is full the front of the pile can be closed in, if it was open before.

A well-managed pile should decompose in one to three months in the summer and three to six months in winter.

A composted pile is hot when it is working and returns to air temperature when it is done. It is crumbly, smells earthy, is dark coloured and looks like potting mix.

Manure use

Ideally, composted manure should be used on the pastures to replace some of the nutrients that grazing horses take out of the system. Unless the property is supporting many horses relative to the acreage (i.e. many horses being kept mainly confined to stables/yards), the amount of manure produced should be able to be spread back onto the paddocks.

Fresh manure should not be spread on pasture currently grazed by horses as it will increase the risk of horse parasite infestation. This risk can be reduced, however, if the paddock is spelled after the manure is spread or is grazed by other types of stock. The amount of time that it takes for the paddock to be ready for horses to graze again, apart from waiting for the grass to reach optimum length, depends on the weather. Hot/dry or very cold periods will kill parasite larvae.

Composted or fresh manure should be applied no more than 2 cm thick on pastures and no more than three to four applications per year. Spread it only during the growing season when it will be used by the plants. Remember that good pasture can utilise compost (or fresh manure) much better than bare soil can. Never spread manure on water-saturated ground and never spread or store it on land subject to flooding.

Manure, either composted or not, should not be used as a surface in high traffic areas (such as for filling in low spots in a training yard for example) because manure holds a lot of water and this area will then become waterlogged when it rains.

If there is too much manure to spread on paddocks (i.e. on a property with a large number of horses in confinement due to lack of pasture), then the surplus can be used as fertiliser on the gardens. If there is still too much it can be sold, traded (i.e. in return for eggs or vegetables) or given away as a last resort.

Fresh, partly or fully composted manure can be marketed to home gardeners, nurseries and crop farmers. Nurseries are the most likely customers for large volumes of less-than-completely composted manure and home gardeners a good outlet for smaller quantities of composted or aged manure.

Fresh manure or compost can be sold at your farm gate. However, in certain areas there are more manure producers (horses) than buyers. This system works best when you are situated in or near suburbia as there will usually be a high demand for manure for gardens in these areas.

Paddock manure management

Manure in the paddock must be managed in some way to reduce the incidence of 'horse sick' pasture. There are various strategies for dealing with paddock manure. What you do with it depends on the size of individual paddocks, the total pasture area on the property and the amount of horses kept on the property.

Collecting and sweeping

The main reason for removing manure from a paddock is to reduce the number of parasitic worms present in the paddock. In larger paddocks collecting manure becomes an ineffective practice. It can reach the stage where more time is spent collecting manure than doing anything else. Some properties go to the extent of buying expensive machinery (sweepers) to do this. However, this still requires an operator and time as in order to be effective (against parasite infestation) manure must be removed either daily or every two days at the most. If manure is left any longer then the benefits of removing it decrease.

Once collected, this manure can be composted and should be returned to the paddocks that need it the most by either spreading by hand or with a machine.

In very small paddocks daily removal of manure is necessary as harrowing becomes more difficult in small spaces and a small paddock will become inundated with manure very quickly. In small, bare paddocks the need to remove manure becomes even more important if it is not to run into the waterway. Avoid keeping horses in bare paddocks due to the land degradation that this will cause.

Spreading manure

In paddocks, manure can be spread by hand with a shovel from the back of a trailer or by a manure spreader. It is now possible to buy small manure spreaders that have been designed for small properties. These machines can be towed behind a ride-on mower, four-wheel bike, utility vehicle or small tractor. If you plan to spread fresh manure daily then the machine can be taken into the stables or yards and manure loaded straight in. If you are composting first, then compost can be put into the machine from the compost bin.

Harrowing

Harrowing gives a quick return of fertiliser (manure) to the soil and causes horses to graze a paddock more evenly by eliminating 'roughs' and 'lawns' (see p. 6). Larger paddocks should be harrowed periodically using pasture harrows. Harrowing involves dragging either a purpose-built pasture harrow or an improvised harrow over the ground with a vehicle (see p. 193).

Dung beetles

Australia and New Zealand both have native dung beetles. However, these are adapted to the manure of native animals, including birds, that have small pellets or piles of manure rather than large piles such as those produced by horses and cows. Some years ago the CSIRO introduced dung beetles that have evolved to deal with such manure into Australia from other countries (such as Europe and Africa). This has been a huge success in Australia. New Zealand also has some introduced species of dung beetles.

There are various types of these beetles and it is usually best to have several types as they tend to work at different times of the year or indeed will work alongside one another for a better result. Dung beetles fly to a fresh pile of manure each evening and usually either burrow down into the soil underneath taking the manure with them or they roll balls of manure to a nearby burrow that they prepared earlier. Either way the result is hugely advantageous in many ways, as shown below.

- By digging tunnels into the ground the soil condition is improved as compacted soil is aerated and rain water can penetrate, this also allows plant roots to penetrate more easily.
- The ground is cleared which enables the grass to grow in that spot straightaway (rather than being trapped under a pile of manure).

Figure 10.3 Dung beetles at work. Notice the fine chopped appearance of the pile. Underneath this pile will be several holes that the bettles have dug in order to take the manure underground. Credit: Jane Myers

- The nutrients in the manure that they take down can be used by the soil immediately as the process of digestion by the beetles makes it more available.
- Fly numbers are reduced as dung beetles remove their habitat.
- Horses will graze in this area rather than reject it (see p. 6) because the manure is taken underground.
- Dung burial also reduces the infective stages of gastrointestinal parasites ('worms') of horses.
- The dung beetles save you an enormous amount of time that would otherwise be taken up with picking up manure or harrowing.
- They create the right conditions for earthworms by loosening the soil. Earthworms are also vital to the soil so it is very important that they be encouraged.

Who would have believed that a small insect could do so much!

Dung beetles will only work on fresh manure that is deposited in the paddock (either via the horse or when fresh manure that is collected from yards and/or stables is spread daily on paddocks). They will not work in a compost bin or on manure that is composted and then spread on paddocks. When dung beetles are working, manure in yards must be picked up daily otherwise that evening the dung beetles take it underground.

See Chapter 12 for information on dung beetle resources.

11

Equipment and tools

It is easy to get carried away and buy, or plan to buy, lots of equipment for a horse property. While some items are undoubtedly invaluable, others will never pay for themselves in terms of how often they are used. For that reason it is usually better to hire certain pieces of equipment and buy others. Buying second-hand or sharing certain items with like-minded friends or neighbours is another option. Sometimes it is more cost effective to use contractors for certain jobs – this will mean that the job is done properly by someone who knows what they are doing.

What equipment is necessary?

Wheelbarrow

On the absolutely essential list is a wheelbarrow. Buy a good strong wheelbarrow as it will be the hardest working piece of equipment on the property.

Brooms and mucking out tools

Also on the essential list would be brooms of varying sizes and tools for picking up manure/mucking out.

Fencing tools

Fencing tools are essential if you plan to make and maintain your own fences. Good tools make the job of erecting a fence much easier, so buy a good pair of fencing pliers, a wire spinner to feed out the wire and a fence strainer that shows the amount of tension on the wire.

Harrows

Another essential piece of equipment is some form of pasture harrow for paddock maintenance. It breaks up and spreads manure and at the same time lightly aerates the top few centimetres of the soil. Pasture harrows can either be commercially made or home-made. Home-made harrows can be made from various materials such as an old bed spring, weldmesh, chainmesh or an old gate. Tyres can be used to add weight. Commercial pasture harrows are usually the most effective as they are flexible (being chain link) and they also have spikes that aerate the soil to some extent. These need either a tractor or at least a large four-wheel drive to pull them around a paddock. Home-made varieties can be made as small as you like – they can be pulled behind a ride-on mower, four-wheel bike or your car.

Other items

Ride-on mower
A ride-on mower can be used for mowing small paddocks. A sturdy model will be required if it is to stand up to much mowing. A small trailer can be attached to the back of a mower which is useful for moving things that are too heavy/awkward to carry. A ride-on mower can also pull some lightweight harrows.

Four-wheel bike
These are commonly purchased for properties but are not as useful as a ride-on mower or garden tractor. On a small property it is hard to justify their use due to the relatively short distances involved.

Garden tractor
A garden tractor is a small version of a full-size tractor. Various implements can be purchased for them making them very useful, if rather expensive.

Tractor
Tractors are very useful but a full-size tractor can cause a lot of compaction. Tractor implements include harrows, front-end loader, slasher/mower, tray, grader blade, fertiliser spreader, post driver.

Chipper/mulcher
A chipper/mulcher machine is useful on a horse property. It can be used to make lucerne chaff from lucerne hay. It can also be used to mulch any clippings from bushes or old hay before putting them in the compost heap.

Whipper snipper
A heavy duty whipper snipper with a blade attachment is very useful as it can be used to slash certain weeds and clumps of grass in the pasture. Whipper snippers are also good for cutting grass that a mower cannot reach, such as under the fence.

Hedge trimmer
As well as trimming hedges, these can be used for cutting branches from fodder trees.

Hand tools
Some hand tools can be very useful such as a chipping tool for weeds, a scythe for cutting long grass and weeds, secateurs for pruning bushes.

Mini manure spreader
These are very useful for spreading manure in the paddock. They can be pulled behind many types of vehicles from a ride-on mower to a tractor.

12

Resources

Recommended further reading

Avery, A. (1997). *Pastures for horses – a winning resource.* RIRDC, Sydney.
Budiansky, S. (1998). *The nature of horses.* Orion Books, London.
Cargill, C. (1999). *Reducing dust in horse stables and transporters.* RIRDC, Canberra.
Cleugh, H. (2003). *Trees for shelter – a guide to using windbreaks on Australian farms.* RIRDC, Sydney.
Chatterton, J. (2000). *John Chatterton's ten commandments.* John and Janet Chatterton, Queensland
De Bairacli-Levy, J. (1991). *The complete herbal handbook for farm and stable.* 4th edn. Faber & Faber, London.
Dixon, P. (Ed.). (1996). *From the ground up: property management planning manual.* Department of Conservation and Natural Resources and Department of Agriculture, Melbourne.
Foyel, J. (1994). *Hoofprints: a manual for horse property management.* Department of Primary Industries, SA.
Fraser, A.F. (1992). *The behaviour of the horse.* Cab International, Oxfordshire.
French, J. (2002). *Natural control of garden pests.* Aird Books, Melbourne.
Hayes, M.H. (1987). *Veterinary notes for horse owners.* 17th edn. Stanley Paul, London.
Hunt, W.F. (1994). *Pastures for horses.* New Zealand Equine Research Foundation, Palmerston North.
Huntington, P., J. Myers & L. Owens (1994). *Horse sense: the guide to horse care in Australia and New Zealand.* 2nd edn. CSIRO Publishing, Melbourne.
Kiley-Worthington, M. (1997). *Equine welfare.* JA Allen, London.
Kohnke, J. (1998). *Feeding and nutrition of horses: the making of a champion.* 3rd edn. Vetsearch International, Australia.
McLean, A. (2003). *The truth about horses: a guide to understanding and training your horse.* Penguin Books, Australia.
Mars, R. & J. Mars. (1998). *Getting started in permaculture.* The Candlelight Trust, WA.

Mills, D. & K. Nankervis. (1999). *Equine behaviour: principles and practice.* Blackwell Science, London.
Mollison, B. (1991). *Introduction to permaculture.* Tagari Publications, NSW.
Nash, D. (1999). *Drought feeding and management for horses.* RIRDC, ACT.
Pollitt, C. (2001). *Equine laminitis/founder.* RIRDC, Sydney.
Rossdale, P.D. & S.M. Wreford. (1993). *The horse's health from A to Z.* David & Charles, Devon.
Rose, R. & M. Offord (Eds). (1997). *Proceedings of the equine nutrition and pastures for horses workshop*, 14–16 Feb 1995, Richmond, Australia. RIRDC, ACT.
Stubbs, A. (1993). *Healthy land, healthy horses – a guidebook for small properties.* RIRDC, Sydney.
Stubbs, A. (1998). *Sustainable land use for depastured horses.* RIRDC, Sydney.
Woodward, P. (1997). *Pest-repellent plants.* Hyland House Publishing, Melbourne.
Yeomans, K.B. (2003). *Water for every farm.* Keyline Designs, Queensland.

Things to do

- Join your local Landcare group where you will find a wealth of information on land care subjects.
- Get together with like-minded neighbours for better buying power, sharing advice and information etc.
- Contact your local TAFE/Technical College and find out what courses they offer that might interest you.

Australia specific resources

Greening Australia: www.greeningaustralia.org.au. This is an organisation that works with landholders, community, government and business to tackle land degradation.

www.gov.au is the portal site for gaining access to all State and Territory governments in Australia. Once at the page of your government, search for land or agencies to find various useful organisations associated with the land.

www.cannylink.com/agricultureaustralian.htm lists lots of sites to do with agriculture.

New Zealand specific resources

The New Zealand Equine Research Foundation is a registered charitable trust that fosters research and education in the New Zealand horse industry. The foundation holds annual seminars, produces books and a quarterly bulletin and has a website. The book *Pastures for Horses* (W.F. Hunt) can be purchased from this site, as well as other relevant books. Their contact details are:
NZ Equine Research Foundation
PO Box 52
Palmerston North
Phone: 64 (0)6 3564940
Email: nzerf@xtra.co.nz
Website: www.nzerf.co.nz

Department of Conservation of New Zealand: www.doc.govt.nz/index.html

Ministry of Agriculture and Fisheries (MAF)
PO Box 2526
Wellington
Phone: 64 (0)4 472 0367
Website: www.maf.govt.nz

The New Zealand Ministry for the Environment: www.mfe.govt.nz

A good site for general information on care of various animals such as horses, sheep, goats, cattle, poultry and that has information on fencing and lots of other interesting articles is www.lifestyleblock.co.nz in the lifestyle file section. This is the site of the magazine *New Zealand Lifestyle Block*. There is also lots of information on weeds that affect New Zealand, however, the advice for the control of these weeds tends to be chemical based.

Other resources

Products

Green Harvest
52/65 Kilcoy Lane
Malany QLD 4552
Phone: 61 (0)7 5494 4676
Email: inquiries@greenharvest.com.au
Website: www.greenharvest.com.au.
This website has numerous interesting articles on organic practices. As well as selling seeds, Green Harvest sells books on such subjects as organic pest control, healthy soils, permaculture and biodynamics. If contacted they will send you a free booklet called *Australian Organic Gardening Resource Guide* that lists the books and other things that they sell and also gives lots of useful information. This is a must-have as it includes natural pest management products, soil test kits, seeds for plants for green manure and fodder such as tagasaste, books on permaculture, biodynamics, animal health (including horse health) and organic gardening.

Soil improvement

The book *Water For Every Farm* by Ken Yeomans has lots of information on soil improvement. Even though the book is aimed at large farms much of the information is relevant to smaller properties. Ken Yeomans also has a website: www.keline.com.au.

Land care

Landcare Australia's website address is: www.landcareaustralia.com.au

Landcare New Zealand's website address is: www.landcare.org.nz

The website of New Zealand's environmental research organisation is: www.landcareresearch.co.nz

A website which has a wealth of information on sustainable property management (much of it horse-based) is: www.landcaresolutions.com.

The horse care magazine *Hoofbeats* has an eight-page section called 'The Green Guide', which provides invaluable information on managing your horse property sustainably. It is available in newsagents, or see www.hoofbeats.com.au

Permaculture

In addition to numerous books on the subject of permaculture (some of which are listed in the recommended further reading list) there are many websites dedicated to the subject. Permaculture started in Australia but is now a worldwide movement. Simply search the word permaculture via your search engine.

Pest control

The company Kingfrog distributes environmentally friendly products for controlling pests. Contact them on 1800 108 888 (Australia) or look up their website www.kingfrog.com.au

Dung beetles

In Australia, John Feehan is a dung beetle consultant who also sells dung beetles:
3 Prell Place
Hackett ACT 2602
Fax/phone (evenings): 61 (0)2 6248 0376
Email: john-feehan@one.net.au

In New Zealand, contact The Ministry of Agriculture and Fisheries (MAF) to find out about the current status of dung beetles:
PO Box 2526
Wellington
Phone: 64 (0)4 472 0367
Website: www.maf.govt.nz

Biodynamic farming

www.kellosheilpark.com is the website for Robert and Catherine McDowell who farm using biodynamics and sell herbs for horses.

See also www.biodynamics.net.au

Organic weed control

Waipuna Systems Ltd: www.waipuna.com. The Waipuna system uses a non-toxic, biodegradable Organic Hot Foam method that is applied to unwanted weeds, killing them instantly. The Organic Foam solution contains natural plant sugar extract from corn and coconut.

Soil testing

An independent (not part of a fertiliser company) laboratory for soil analysis is SWEP Analytical Laboratory:
47/174 Bridge Rd
Keysborough Vic 3173
Phone: 61 (0) 3 9701 6007
Email: services@swep.com.au
Website: www.swep.com.au
SWEP analyse soils and plant tissue and are also used by clients from other countries, including New Zealand.

Horse care and health

The website www.equinecentre.com.au has a wealth of information on horse care and horse health.

Rural Industries Research and Development (RIRDC) books can be purchased from RIRDC
Phone: 61 (0) 2 6272 4819
Email: publications@rirdc.gov.au
Website: www.rirdc.gov.au/eshop. Some RIRDC books are also available as free downloads from this site.

Vaccinations

Have a look at the Commonwealth Serum Laboratories website: www.csl.com.au for more information on strangles, tetanus and their vaccinations.

Hoof care and bare foot trimming

David Farmilo, a master farrier, has an excellent website on hoof care:
http://www.farmilo.com.au/horse.htm
See also: www.strasserhoofcareaustralia.com and www.naturalhoof.co.nz.

Fire and flood

www.equinecentre.com.au has fact sheets that can be downloaded on the subjects of fire and flood.

Water

Australian Water Association: www.awa.asn.au. This site gives details on the responsible management of water and related resources.

www.mfe.govt.nz is the website of the New Zealand Ministry for the Environment. Go to the page on water for information on water management.

Animal movement in Queensland

www.dpi.qld.gov.au is a website that gives information on stock movements in and out of tick infested areas, local diseases and animal management in Queensland.

Drought management

The book *Drought Feeding and Management for Horses* by David Nash is a must have for horse owners that may be affected by drought, and is published by RIRDC, ACT. This book can be purchased or downloaded from their website at the publications page of their website: www.rirdc.gov.au/eshop.

Horse handling

The book *John Chatterton's Ten Commandments* is an excellent guide to safe horse handling. It can be purchased in some saddlery stores, or direct from the author: John Chatterton, phone 61 (0)7 5546 3146.

Trees and plants

An edited version of the book, *Trees for Shelter – A Guide to using Windbreaks on Australian Farms*, by H. Cleugh can be downloaded free from the RIRDC website: www.rirdc.gov.au/reports/AFT/02-162sum.html.

Identifying plants

www.plant.id.au is a site dedicated to plant identification

Poisonous plants

The Australian Government Department of the Environment and Heritage, National Botanic Gardens website: www.anbg.gov.au/bibliography/poison-plants.html has a list of poisonous plants in Australia. The site also lists books on plants that are poisonous to animals and plants that affect allergies and dermatitis in humans.

The Manaaki Whenua Landcare website: www.landcareresearch.co.nz/publications/infosheets has information on poisonous plants in New Zealand.

Glossary

abscess: a cavity containing dead cells, bacteria and pus.
acidic: soil that has a pH of less than seven in water.
ad lib: ad libitum: no limit.
agistment: to provide feed, shelter and water for horses for payment.
agronomist: person trained in the area of soil management and crops.
alkaline: soil that has a pH of more than seven in water.
annual: plants that flower and die in one year.
arena: an enclosed area, usually rectangular, used for exercising a horse.
biannuals: plants that live for two years.
biodiversity: a diverse ecosystem.
biodynamics: a form of farming using natural fertilisers rather than chemical fertilisers.
block grazing: strip grazing with second electric fence to stop the animals from going back over the grazed area.
bot: botfly larva which lives in the horse's stomach. Also refers to the free-living fly stage.
box: a stable for a single horse.
burrs: seed pods covered in spines.
cattle grid: a series of metal bars with gaps between designed to discourage cattle from crossing through an open gateway.
certified seed: seed that is certified as having a minimum purity and germ and of a particular variety.
chaff: hay or straw cut into short lengths for use as a feedstuff.
colic: pain in the abdomen.
condition scoring: a method of estimating the condition of a horse.
concentrates: grains etc. that make up the non-roughage part of the feed.
continuously grazed: paddocks that are grazed without resting.
crest: the upper line of a horse's neck.
cribbing: stereotypic behaviour in which the horse grasps wooden objects with its front teeth and swallows air.
cross-grazing: the practice of using different animal species to graze a paddock.
CSIRO: Commonwealth Scientific and Industrial Research Organisation (Australia).
cultivar: a selectively bred variation of a plant species.
dam: female parent of a horse and water storage pond.
dander: particles of skin (or feathers) shed from an animal or bird that may act as allergens.
deep tillage: deep ripping of the soil with machinery to aerate it.

digestibility: the percentage of the feed eaten by the horse that the horse can actually utilise.
direct drilling: grass seed is put into the soil by machine rather than simply broadcast over the top.
DM: dry matter (in food).
dormant: the stage of a plant's cycle when it is unable to grow.
dressage: the art of training horses to perform all movements in a balanced, supple, obedient and keen manner.
dry sheep equivalent (DSE): basic unit of measurement that is used to calculate stocking rates. It is based on a 45kg merino wether.
electrolytes: minerals Sodium, Potassium Chloride and Bicarbonate that are important in maintaining hydration and pH status.
endophyte: a fungus that lives within grasses and provides benefits to them, however, they can cause problems with animal health.
energy: the horse gets energy from feed.
equine: horses are part of the family *equidae* which also includes asses and zebras.
farrier: person who trims hooves and fits shoes.
fertiliser: material that is added to the soil to supply chemical elements needed for plant nutrition.
float: a special trailer for carrying one or more horses.
fore-gut fermenter: cattle and most other grazing animals (apart from horses) are termed fore-gut fermenters because most of the digestion of feed takes place in the stomach.
forage: to forage is to graze. The term forage is also applied to any type of feed for stock
foal: a young horse up to the age of 12 months.
founder: see laminitis.
fibre: the fibrous substance of plants.
fungicide: an agent that kills or destroys fungi.
good doer: a horse that gains weight or stays fat on minimal feed.
greasy heel: skin infection, usually on coronet, heels and pasterns. Also referred to as mud fever.
gypsum: a calcium sulphate mineral used to improve soil structure.
hand: a linear measurement equalling 4 inches or 10 centimetres, used in giving the height of a horse from the ground to its withers.
hard feed: term for concentrates fed to horses, i.e. grain or mixes.
head collar: a bitless headpiece and noseband, usually of leather or nylon, for leading a horse that is not wearing a bridle.
herbage: herbaceous growth or vegetation such as pasture.
hind-gut fermenter: horse are hindgut fermenters because they digest most of their food in the hind gut *after* the stomach unlike fore gut fermenters (*see* fore-gut fermenters).
horse sick: used to describe pasture that has rank grass areas containing lots of dung.
humus: decomposed organic matter in the soil.
inorganic: not having the structure or characteristics of living organisms.
immunity: the body's natural response to challenge from foreign material such as bacteria, viruses, cells.
land degradation: general term for various conditions resulting in damaged land.
laminitis/founder: inflammation of the sensitive laminae of the hoof, characterised by heat and pain.
lawns: the areas that the horse grazes low to the ground in a horse sick pasture.
leach: to take away or drain nutrients from the soil.
legume: a type of plant (i.e. clover and lucerne) that provides high protein feed. It has the ability to convert nitrogen in the atmosphere to a form that plants can use.
lime: a product used to treat acid soils. It is made from limestone.
lunge: to exercise a horse on the end of a rope in a circle.
macrominerals: minerals needed in relatively large amounts.

manure: faeces passed by horses, also called droppings.
microminerals: minerals needed in trace amounts only.
minerals: naturally occurring substances with characteristic chemical composition.
muck out: to clean out a stable of droppings and dirty wet bedding.
mud fever: skin infection of heels and pasterns. See also greasy heel.
mulching: 1. chopping up organic material into small pieces so that it breaks down more quickly. 2. putting mulched material around plants to protect them from temperature extremes and to hold in moisture.
nitrate leaching: the draining or taking away of nitrates in the soil profile.
organic: produced by, or found in plants and animals.
organic matter: the matter found in soils which is from plants or animals.
over-sowing: reseeding pasture with seeds broadcast on top of undisturbed soil.
overgrazing: not allowing pasture plants to recover from grazing pressure leading to loss of desired species and increase in undesirable species (weeds).
oxalates: organic compounds based on oxalic acid.
palatability: this describes how much an animal prefers a certain feed type compared to another.
pasture sweeping: removal of weeds, manure and leaves from a paddock by machine.
pasture topping: slashing of pasture to cut down rank grass and weeds.
parasite: an organism that lives on or inside the horse.
permaculture: a system of living that utilises the environment but enhances it rather than takes from it.
pony: usually a horse of any breed up to 14.2 hands.
poor doer: a horse that loses or fails to gain weight with normal feed supply.
perennial: a plant that lives for many years, sets seed each year.
pugging: the damage caused by animals' hooves in wet soils.
protein: a class of chemical formed of amino acids essential to all living things.
resident grasses: the permanent grasses in a paddock that have persisted for a number of years.
ripping: deep tillage of the soil.
RIRDC: Rural Industries Research and Development Corporation (Australia).
roughage: high fibre feed such as pasture, hay, chaff.
rhizomes: a thick horizontal underground stem whose buds develop into new plants.
rotational grazing: the practice of moving stock around paddocks in order to rest paddocks after grazing.
roughs: tall rank grass in the areas that horses manure in the paddock.
ruminant: an animal that regurgitates and re-chews its food (cattle, sheep etc.) (also called a fore-gut fermenter) as opposed to a hind-gut fermenter (e.g. equines).
rural urban fringe: the area on the edge of the city that is a mix of urban and rural land.
sacrifice area: part of a property that is fenced off and used to confine animals until pasture is in the optimum state to stand grazing.
sand roll: an area of sand for rolling in after a horse is worked or washed.
scouring: diarrhoea, passing of loose watery faeces.
seedy toe: the separation of sensitive and insensitive laminae of the foot. Follows chronic laminitis/founder or hoof cracking.
silage: fodder that has been fermented to conserve it.
soil compaction: process whereby soils lose air space due to heavy weight being asserted on the ground level.
soil structure: the arrangement of the soil particles and the pore space around them.
soundness: state of health or fitness in horses to carry out a particular function.
species: a distinct class, sort, or kind, of something (plants and animals).

stable: either a building for keeping horses or an individual box for a horse.
staring coat: a coat standing up and looking dull.
stocking rate: the number of stock in a certain area.
strangles: an infectious and highly contagious disease caused by the bacteria *Streptoccus equi.*
strain: a natural variation of a plant species.
stringhalt: involuntary snatching up of the hind leg and flexing of the hock when walking.
stolons: stem-like parts of plants that grow along the ground surface and from which new buds arise.
strip grazing: using temporary electric fences to allow animals to gradually graze across a paddock rather than have access to the whole paddock.
subsoil: the layer of soil that is immediately below the topsoil.
supplement: extra feed other than pasture.
sward: the leaf of a grass plant.
tack: saddlery.
tetanus: an often fatal disease related to wounds.
thrush: inflammation of the frog of a horse's foot, characterised by a foul-smelling discharge.
tiller: an individual shoot of a grass plant.
topdressing: applying fertiliser or manure to the surface of the soil.
topping: slashing or mowing the excess growth in a paddock.
under sowing/drilling: planting new seed under the surface of an existing pasture with a seed drill.
urban fringe: the area that surrounds the suburbs of a city.
vaccinate: to inject vaccine to stimulate immunity, e.g. for tetanus.
vegetative stage: the stage of growth before the plant sets seed.
vice: a common term for horses that display stereotypic behaviours (e.g. windsucking).
waterlogged soil: the soil is compacted and saturated to the extent that the air spaces are forced out.
weed: any plant that is in the wrong place at the wrong time.
weaving: stereotypic behaviour characterised by the horse rocking from side to side, usually over a stable door or gate.
windsucking: a stereotypic behaviour that is characterised by the horse gulping and swallowing air.

Bibliography

Andrews, G.J. (1998). The use of trees and shrubs for livestock production. *Australian Biologist* 11(2).

Anon. (2002). Plastic fencing. In: 'The Green Horse', *Hoofbeats Magazine* October/November.

Avery, A. (1997). *Pastures for horses – a winning resource.* RIRDC, Sydney.

Baylis, K. (1999). Soil testing – the first step. In: 'The Green Horse', *Hoofbeats Magazine* April/May 1999.

Blickle, A.R. (2001) Healthy Horses, Clean Water: Horses for clean water: A guide to Environmentally Friendly Horse Keeping. Alayne Renee Blickle, Seattle, WA.

Brookman, P. (1999). A happy medium. In: 'The Green Horse', *Hoofbeats Magazine* June/July.

Cargill, C. (1999). *Reducing dust in horse stables and transporters.* RIRDC, Canberra.

Cleugh, H. (2003). *Trees for shelter – a guide to using windbreaks on Australian farms.* RIRDC, Sydney.

Dixon, P. (Ed.). (1996). *From the ground up: property management planning manual.* Department of Conservation and Natural Resources and Department of Agriculture, Melbourne.

Ehringer, G. (1995). *Roofs and rails – how to plan and build your ideal horse facility.* Western Horseman Inc., USA.

Ferreira, C. & T. Bell. (2003). Making your farm fire safe. In: 'The Green Horse ', *Hoofbeats Magazine* April/May.

Fereira, C. & T. Bell. (1999). Buying an equestrian property. *Hoofbeats Magazine* February/March.

Fereira, C. & T. Bell. (1999). Property hunting. In: 'The Green Horse', *Hoofbeats Magazine* June/July.

Foyel, J. (1994). *Hoofprints: a manual for horse property management.* Primary Industries, SA.

Gordon, J. (2001). *The horse industry – contributing to the Australian economy.* RIRDC, Canberra.

Hill, C. (1990). *Horsekeeping on a small acreage – facilities design and management.* Storey Communications, USA.

Hunt, W.F. (1994). *Pastures for horses.* New Zealand Equine Research Foundation, New Zealand

Huntington, P., J. Myers, & L. Owens (1994). *Horse sense: the guide to horse care in Australia and New Zealand.* 2nd edn. CSIRO Publishing, Melbourne.

Kohnke, J. (1998). *Feeding and nutrition of horses: the making of a champion.* 3rd edn. Vetsearch International, Australia.

Pluske, W. (1998). What is soil anyway? In: 'The Green Horse' *Hoofbeats Magazine* Febuary/March.

Pluske, W. (1998). Soils have structure. In: 'The Green Horse' *Hoofbeats Magazine* April/May.

Rose, R. & M. Offord (Eds). (1997). *Proceedings of the equine nutrition and pastures for horses workshop*, 14–16 Feb 1995, Richmond, Australia. RIRDC, ACT.

Stubbs, A. (1993). *Healthy land, healthy horses – a guidebook for small properties.* RIRDC, Sydney.

Stubbs, A. (1998). *Sustainable land use for depastured horses.* RIRDC, Sydney.

Warren, L.K. & C. Sweet. (2003). *Manure and pasture management for horse owners.* AAFRC, Canada .

West, G. (Ed.). (1988). *Blacks veterinary dictionary.* A&C Black, London.

Yeomans, K.B. (2003). *Water for every farm.* Keyline Designs, Queensland.

Index